Leckie
the education publisher
for Scotland

National 4 & 5
GEOGRAPHY

Course Notes

Patricia Coffey

ISBN 9780008282127

Published by
Leckie
An imprint of HarperCollinsPublishers
Westerhill Road, Bishopbriggs, Glasgow, G64 2QT
T: 0844 576 8126 F: 0844 576 8131
leckiescotland@harpercollins.co.uk www.leckiescotland.co.uk

HarperCollins Publishers
Macken House, 39/40 Mayor Street Upper, Dublin 1, D01 C9W8, Ireland

Special thanks to
Roda Morrison (project management and copyedit)
Rona Gloag (proofread)
Jouve (layout and illustrations)
Printed in the UK

A CIP Catalogue record for this book is available from the British Library.

This book contains FSC™ certified paper and other controlled
sources to ensure responsible forest management.

For more information visit: www.harpercollins.co.uk/green

Acknowledgements

We would like to thank the following for permission to reproduce their material:
Photography/Shutterstock.com; Photograph of the middle course of the River Clyde on page 70 © Mary and Angus Hogg;
Photograph of Battlefield Cavern on page 87 © Duncan Jones; Photograph of stalactites and stalagmites on page 88 © Jason Patrick
Ross/Shutterstock.com; Photograph of caving on page 92 © Barcroft Media via Getty Images; Photograph of the Ingelton Waterfalls
Trail on page 94 © Alasdair Milne; Photograph of Loch Einich on page 110 © Angus; Photograph of Cairngorm Funicular Railway
on page 110 © Paul Birrel; Photograph of hill walkers on page 111 © Roger Antrobus; Photograph of Weymouth Beach on page
129 © BasPhoto/Shutterstock.com; Photograph of A351 on page 129 © Ian Britton/FreeFoto; Photograph of windsurfing on page
130 © EML/Shutterstock.com; Photograph of Poole Harbour on page 130 © VisitBritain/Rod Edwards; Photograph on pages 140–1
© am70/Shutterstock.com; Photograph of city on page 145 © komar/shutterstock.com; Photograph of rural area on page 145 ©
Giancarlo Liguori/Shutterstock. com; Photograph of desert on page 146 © oksana.perkins/Shutterstock.com; Photograph of
mountainous region on page 146 © Hildegard Willer; Photograph of rainforest on page 147 © Andre Nantel/Shutterstock.com;
Photograph of polar region on page 147 © Vitalii Nesterchuk/Shutterstock.com; Photograph of Glasgow CBD on page 163 ©
cornfield / Shutterstock.com; Photograph of the demolition of old tenements on page 173 © Glasgow Digital Library, based at the
University of Strathclyde; Photograph of new housing development on page 174 © Thomas Nugent; Photograph of Hillington
Industrial Estate on page 178 © The Robert Pool Glasgow Collection; Photograph of intu Braehead on page 179 kay roxby /
Shutterstock.com; Photograph of police in Roçhina on page 182 © AFP/Getty Images; Photograph of rubbish on the street of the
favela on page 183 © LatinContent/Getty Images; Photograph of Roçinha on page 184 © Matyas Rehak / Shutterstock.com;
Photograph of a charity-funded school on page 186 © Developing Minds Foundation; Photograph of tractor on page 198 © Taina
Sohlman/Shutterstock.com; Photograph of rice fields on page 202 © gallimaufry/Shutterstock.com; Photograph on pages 212–3 ©
Stefano Guzzetti/Shutterstock.com; Photograph of gold mining on page 219 © Ryan M. Bolton/Shutterstock.com; Photograph of an
Inuit hunter on page 223 © David Hiser; Photograph of Fukushima on page 238 © Propaganda/Alamy; Photograph of Sendai
Airport © Samuel Morse; Photograph of seismograph on page 239 © jamesbenet; Photograph of Soufriere Hills on page 243 ©
Justin Kasez10z/Alamy; Photograph of Plymouth on page 244 © Andrew Woodley/Alamy; Photograph of satellite image of
Hurricane Sandy on page 248 © Getty Images; Photograph of damaged property in New York on page 248 © Leonard Zhukovsky/
Shutterstock.com; Photograph of Times Square on page 258 © S.Borisov/Shutterstock.com; Photograph of Spanish beach on page
259 © nito/Shutterstock.com; Photograph of river tour of Borneo on page 263 © Andrew Watson; Photograph of tourist lodge in
Borneo on page 265 © Jane Sweeney; Photograph of Orang-utan on page 265 © Eric Gevaret/Shutterstock.com

All other images from Shutterstock.com

OS Maps
Ordnance Survey ® map data licensed with the permission of the Controller of Her Majesty's Stationery Office.
© Crown copyright. Licence number 100018598.

National 5 exam style questions
Sample answers to these exam-style questions can be found on the Leckie website at
https://collins.co.uk/pages/scottish-curriculum-free-resources

Whilst every effort has been made to trace the copyright holders, in cases where this has been unsuccessful, or if any have
inadvertently been overlooked, the Publishers would gladly receive any information enabling them to rectify any error or omission
at the first opportunity.

Ordnance Survey

OS Landranger Map / 1:50 000 Scale Colour Raster

Customer Information

Additional data sourced from third parties, including public sector information licensed under the Open Government Licence v1.0

Whilst we have endeavoured to ensure that the information in this product is accurate, we cannot guarantee that it is free from errors and omissions, in particular in relation to information sourced from third parties

Reproduction in whole or in part by any means is prohibited without the prior written permission of Ordnance Survey

Ordnance Survey, the OS Symbol, OS and Landranger are registered trademarks of Ordnance Survey, the national mapping agency of Great Britain

Communications

PUBLIC RIGHTS OF WAY

DROIT DE PASSAGE PUBLIC   ÖFFENTLICHE WEGERECHTE

—————— Footpath

——————— Restricted byway (not for use by mechanically propelled vehicles)

—·—·—·—·— Bridleway

+·+·+·+·+·+ Byway open to all traffic

Public rights of way shown on this map have been taken from local authority definitive maps and later amendments

The symbols show the defined route so far as the scale of mapping will allow.
Rights of way are not shown on maps of Scotland

Rights of way are liable to change and may not be clearly defined on the ground. Please check with the relevant local authority for the latest information

The representation on this map of any other road, track or path is no evidence of the existence of a right of way

OTHER PUBLIC ACCESS

AUTRES ACCÈS PUBLICS   ANDERE ÖFFENTLICHE WEGE

· · · Other route with public access {not normally shown in urban areas

♦ National Trail, European Long Distance Path, Long Distance Route, selected Recreational Routes

The exact nature of the rights on these routes and the existence of any restrictions may be checked with the local highway authority. Alignments are based on the best information available. These routes are not shown on maps of Scotland

● ● ● On-road cycle route
○ ○ ○ Traffic-free cycle route

■4■ National Cycle Network number
■8■ Regional Cycle Network number

Danger Area   Firing and Test Ranges in area. Danger! Observe warning notices.
Champs de tir et d'essai. Danger! Se conformer aux avertissements.
Schiess-und Erprobungsgelände. Gefahr! Warnschilder beachten.

ROADS AND PATHS

VOIES DE COMMUNICATION   STRASSEN UND WEGE

Not necessarily rights of way

Service area Ⓢ | Motorway (dual carriageway)
Autoroute (chaussées séparées) avec aire de service et échangeur numéroté
Autobahn (zweibahnig) mit Servicestation und Anschlussstelle sowie Nummer der Anschlussstelle

M1 Elevated / En Viaduc / Erhöht
Junction number

| Unfenced | Dual carriageway |
| Sans clôture | Chaussées séparées / Zweibahnige Strasse |
A 470 | Primary Route / Itinéraire principal / Fernstrasse

A 493 | Footbridge / Passerelle / Fussgängerbrücke
Main road / Route principale / Hauptstrasse

Road under construction / Route en construction / Strasse im Bau

Nicht eingezäunt
B 4518 | Secondary road / Route secondaire / Nebenstrasse

A 855 | B 885
Bridge / Pont / Brücke

Narrow road with passing places
Route étroite avec voies de dépassement
Enge Strasse mit Ausweichstellen

Road generally more than 4m wide
Route généralement de plus de 4m de largeur
Strasse, im allg.über 4m breit

Road generally less than 4m wide
Route généralement de moins de 4m de largeur
Strasse, im allg.unter 4m breit

Other road, drive or track
Autre route, allée ou sentier
Sonstige Strasse, Zufahrt oder Feldweg

Path / Sentier / Fussweg

Gradient: steeper than 20% (1 in 5)   14% to 20% (1 in 7 to 1 in 5)
Pente: Supérieure à 20% (1 pour 5)   14% à 20% (1 pour 7 à 1 pour 5)
Steigung über 20%   14% bis 20%

Gates / Barrières / Schranken
Road tunnel / Tunnel routier / Strassentunnel

Ferry (passenger) / Bac pour piétons / Personenfähre
Ferry (vehicle) / Bac pour véhicules / Autofähre

Ferry P   Ferry V

PRIMARY ROUTES

These form a network of recommended through routes which complement the motorway system

RAILWAYS

CHEMINS DE FER   EISENBAHNEN

—————— Track multiple or single

- - - - - Track under construction

Light rapid transit system, narrow gauge or tramway

Bridges, footbridge

Tunnel, cutting

a Station, (a) principal

Siding

Light rapid transit system station

LC Level crossing

Viaduct, embankment

Ordnance Survey®

General Information

LAND FEATURES

Electricity transmission line (pylons shown at standard spacing)

Pipe line (arrow indicates direction of flow)

Buildings

Important building (selected)

Bus or coach station

Glass structure

Heliport

Current or former place of worship with tower

Current or former place of worship with spire, minaret or dome

Place of worship

Triangulation pillar

Mast

Wind pump, wind turbine

Windmill with or without sails

Graticule intersection at 5' intervals

Cutting, embankment

Landfill site or slag/spoil heap

Coniferous wood

Non-coniferous wood

Mixed wood

Orchard

Park or ornamental ground

Forestry Commission land

National Trust-always open

National Trust-limited access, observe local signs

National Trust for Scotland-always open

National Trust for Scotland-limited access, observe local signs

Manx National Heritage

Isle of Man Forestry Division Plantation

BOUNDARIES

National

District

County, Unitary Authority, Metropolitan District or London Borough

National Park

WATER FEATURES

Marsh or salting

Towpath

Aqueduct

Weir

Footbridge

Lock

Bridge

Ford

Normal tidal limit

Canal (dry)

Contour values in lakes are in metres

Cliff

Slopes

Sand

Dunes

High water mark

Low water mark

Flat rock

Lighthouse (in use)

Lighthouse (disused)

Beacon

Mud

Shingle

HEIGHTS

Contours are at 10 metres vertical interval

Heights are to the nearest metre above mean sea level

Where two heights are shown, the first height is to the base of the triangulation pillar and the second (in brackets) to the highest natural point of the hill

---

ABBREVIATIONS

More information on abbreviations and symbols can be found on our website

CH	Clubhouse	CG	Cattle grid
PH	Public house	P	Post office
PC	Public convenience (in rural area)	MP	Milepost
TH	Town hall, Guildhall or equivalent	MS	Milestone

ARCHAEOLOGICAL AND HISTORICAL INFORMATION

+ Site of antiquity   VILLA Roman   ✕ Battlefield (with date)

☆ ···· Visible earthwork   Castle Non-Roman

Information provided by English Heritage for England and the Royal Commissions on the Ancient and Historical Monuments for Scotland and Wales

ROCK FEATURES

Outcrop

Cliff

Scree

Tourist Information

TOURIST INFORMATION   RENSEIGNEMENTS TOURISTIQUES   TOURISTENINFORMATION

Selected places of tourist interest
Endroits d'un intérêt touristique particulier
Ausgewählter Platz von touristischem Interesse

Telephone, public / roadside assistance
Téléphone, public / borne d'appel d'urgence
Telefon, öffentlich / Notrufsäule

Camp site/caravan site
Terrain de camping/Terrain pour caravanes
Campingplatz/Wohnwagenplatz

Garden
Jardin
Garten

Golf course or links
Terrain de golf
Golfplatz

Information centre, all year / seasonal
Office de tourisme, ouvert toute l'année / en saison
Informationsbüro, ganzjährig / saisonal

Nature reserve
Réserve naturelle
Naturschutzgebiet

Parking / Park & Ride, all year / seasonal
Parking / Parking et navette, ouvert toute l'année / en saison
Parkplatz / Park & Ride, ganzjährig / saisonal

Picnic site
Emplacement de pique-nique
Picknickplatz

Viewpoint
Point de vue
Aussichtspunkt

Visitor centre
Centre pour visiteurs
Besucherzentrum

Walks / Trails
Promenades
Wanderwege

Youth hostel
Auberge de jeunesse
Jugendherberge

World Heritage site/area
Site du Patrimoine Mondial
Welterbestätte

Recreation / leisure / sports centre
Centre de détente / loisirs / sports
Erholungs- / Freizeit- / Sportzentrum

CONVERSION

METRES - FEET

1 metre = 3.2808 feet

Metres 0

15.24 metres = 50 feet

Feet 0

It is important to study Geography as it enables us to develop a number of different skills. Geographical skills include mapping, field work, research and, interpreting and presenting numerical and geographical information. Geographers are also able to further develop their literacy and numeracy competencies within a variety of contexts. Thinking skills are deepened in a wide range of geographical settings. Learners can enhance their abilities in employability, enterprise and citizenship. These skills are transferrable and allow Geographers to develop skills for learning, life and work.

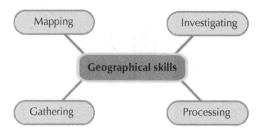

Consider the different subjects where your Geographical skills are transferrable.

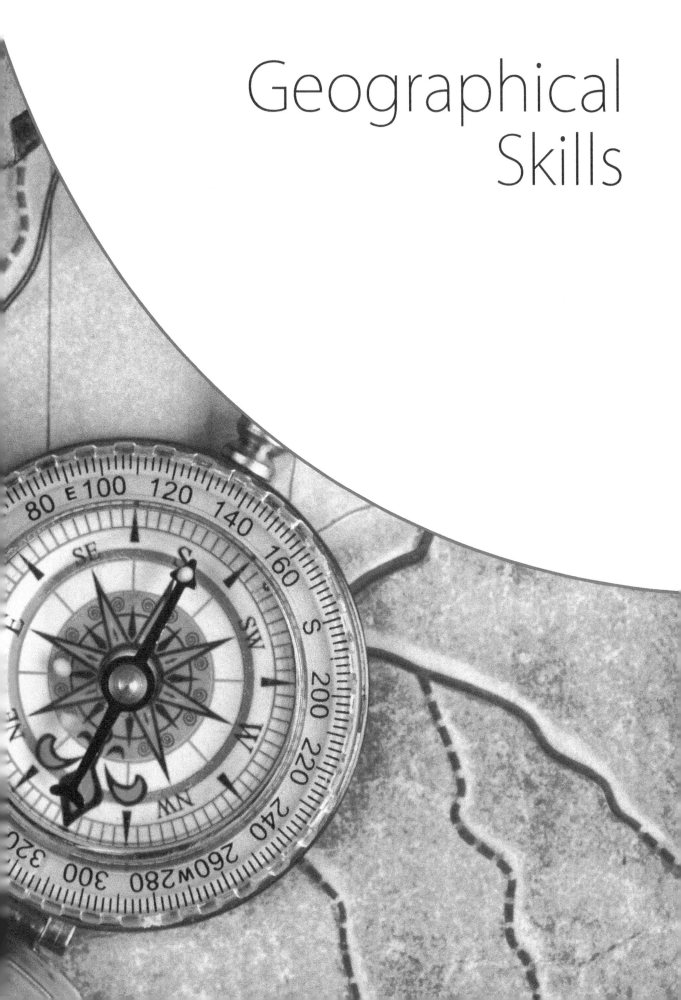

Geographical
Skills

1 Geographical skills

Within the context of geographical skills, you should know and understand:

- Techniques for gathering geographical information.
- Methods of processing data collected in Geography.
- Strategies for researching various aspects of the subject including physical and human geography and global issues.

You also need to develop the following skills:

- Mapping skills related to Ordnance Survey maps including: 4 and 6 figure grid references, measure distance using scale, interpret relief and contour patterns, use maps in association with photographs, field sketches, cross sections and transects.
- Extracting, interpreting and presenting numerical and graphical information which may be: graphs, tables, diagrams or maps.

Map skills

Direction

A **compass** is used to give directions. On most maps north is at the top but, before you begin to give directions, remember to check where north is on your map by looking for the compass.

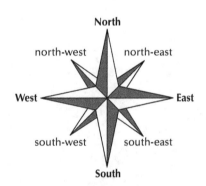

Figure 1.1: *An eight-point compass*

Scale

The scale of a map tells you what the real distance is on the ground. For the National 5 exam, you should be familiar with working out the distance between different points on both 1:25 000 and 1:50 000 **Ordnance Survey (OS)** maps. The easiest way to calculate a straight line distance on an OS map is to use a ruler to measure the distance on the map in centimetres. Next, use the scale line at the bottom of the map to calculate the real distance. Always remember to give your answer in either metres or kilometres:

- On a 1:25 000 map, 4 cm on the map is equal to 1 km on the ground.
- On a 1:50 000 map, 2 cm on the map represents 1 km on the ground.

🔍 **HINT**

To help you learn the positions of **W**est and **E**ast on a compass, remember they spell out the word '**WE**'.

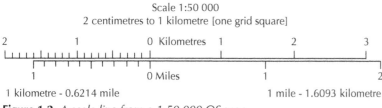

Figure 1.2: *A scale line from a 1:50 000 OS map*

Grid references

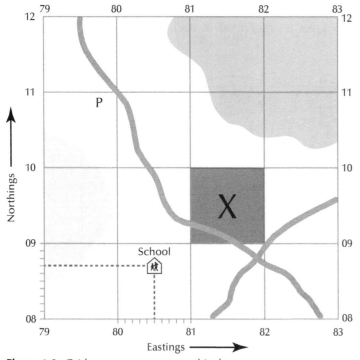

Figure 1.3: *Grid system on a topographical map*

OS maps use a grid system. The grid lines that run from north to south are called **eastings** and those which run from west to east are called **northings**. The grid lines are used to locate and give grid references for different symbols or features on an OS map. 4-figure grid references allow you to locate a grid square and 6-figure grid references allow you to pinpoint a specific symbol. In Figure 1.3, the 4-figure grid reference for the square marked X is 8109. The 6-figure grid reference for the school is 805087.

Relief

Relief refers to the height of the land and it can be shown on an OS map in different ways:

1. A **spot height**, e.g. •985. This means that the highest point on the map is 985 metres at that grid reference.

🔵 Make the Link

In the Rivers section, you should be able to give the direction from which the river is flowing on an OS map.

🔵 Make the Link

In the Weather chapter, you should know that the direction of the passage of a depression is from west to east across the UK.

🔍 HINT

An easy way to convert centimetres to metres is to divide the number by 100. To convert centimetres to kilometres, divide by 1000.

🔍 HINT

Remember to use the key on an OS map to help you identify symbols.

🔵 Make the Link

In the physical environments and human environments chapters, you should be able to give both 4- and 6-figure grid references from an OS map.

🔍 HINT

In the Weather chapter, black lines with numbers on them are called **isobars** and join areas of equal air pressure – be careful not to confuse them with contour lines!

> ### :: Make the Link
>
> In the Glaciated uplands chapter, when identifying a pyramidal peak you should look for a spot height as it marks the highest point.

> ### :: Make the Link
>
> In Maths, gradient is the term used to describe the steepness of a slope.

> ### :: Make the Link
>
> Figure 1.4 also shows different river features including a **meander**, **flood plain**, **confluence** and **estuary**.

2. A **triangulation pillar** or trig point. It is shown on an OS map like this: △ 525. On the ground there is usually a concrete pillar marking the highest point in metres.

3. Brown/orange lines with numbers on them called **contour lines**. When contour lines are very close, the land is steep. When there are no contours and just white space, the land is said to be flat.

Landforms

Contour lines also show the shape of the land. Figure 1.4 illustrates some different **contour patterns** that represent various landforms. Figure 1.5 gives a description of each landscape feature shown on the map.

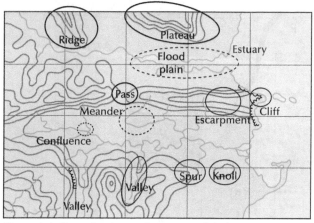

Figure 1.4: *Some landscape features and their contour patterns*

Type of landscape feature	Description
Cliff	very steep (vertical) slope
Escarpment	a ridge with one side that is much steeper (scarp slope) than the other (dip slope)
Knoll	small, isolated hill
Pass (gap)	long valley providing a natural routeway through an upland area
Plateau	flat-topped area of high land
Ridge	long, narrow area of higher land
Spur	finger of high land that juts out into an area of lower land
Valley	narrow area of low land that has higher land on both sides; valleys widen as they approach the sea

Figure 1.5: *Descriptions of landscape features*

> ### GO! Activity 1 (National 5)
>
> Individually, collect an OS map from your teacher. Make up a question about (a) to (h) below and work out the answers to test your map skills.
>
> **(a)** Direction
>
> **(b)** Distance and scale
>
> **(c)** 4-figure grid references
>
> **(d)** 6-figure grid references
>
> **(e)** Spot heights
>
> **(f)** Triangulation pillars
>
> **(g)** Contour lines referring to height
>
> **(h)** Contour lines related to different landforms.

Gathering geographical information

Field work is an essential aspect of Geography. It not only allows you to collect information first-hand in the field but also provides you with many transferrable skills. There are various ways in which you can gather information and some of these are outlined in the sections that follow.

Extracting information

Ordnance Survey (OS) maps are a very common source of geographical information as maps provide us with a wealth of evidence. You can also extract information from other sources including books, websites, TV programmes and news and weather reports. In the physical landscapes chapters (Rivers and valleys, Upland limestone, Coastal landscapes and Glaciated uplands) you will learn how to describe various physical features from OS maps including land height, slope and landforms. In the Urban chapter you will learn how to describe different land use zones on an OS map. Comparing old and new maps is a good way to show changes over time, e.g. how land use has changed in a city over the years.

> ## 🔍 HINT
> "Digimaps" is a great software package for accessing OS maps.

Observing and recording

These two **gathering techniques** often go together because when you are gathering information you have to note down things that you have seen. Aspects of Geography that can be observed and recorded include cloud amount, pedestrians, traffic, land use and environmental quality. A **traffic survey** record sheet (Figure 1.6) can be used to gather information on the number of vehicles in an area. An **environmental quality survey** (Figure 1.7) is used to record the quality of aspects of the environment.

TRAFFIC SURVEY RECORD SHEET

Date: Location:

Start time: Finish time:

Task: Stand on the pavement back from the road. Observe the type of vehicles that pass by you and record them using tally marks, in the table below:

cars	buses	lorries	vans	motorbikes	bikes

Figure 1.6: *A traffic survey sheet*

ENVIRONMENTAL QUALITY SURVEY
Date: Time: Location:
Task: Survey your area and give it a mark out of 10 for each aspect of environmental quality. Record your observations in the table below:

Low quality (0)	Points	High quality (10)
Badly maintained buildings		Clean, attractive buildings
Ugly surroundings		Attractive surroundings, e.g. park
A lot of traffic		Little traffic
Noisy		Quiet
A lot of graffiti and vandalism		No graffiti or vandalism
Air pollution		Clean air
No greenery visible		Lots of trees and grass
A lot of derelict land/waste ground		No derelict land/waste ground
A lot of litter		No litter
A lot of dog dirt		No dog dirt
	TOTAL =	

Figure 1.7: *An environmental quality survey*

Measuring

When measuring different features of the physical and human environments in Geography, special instruments are used such as:

Weather	Rivers	Urban
A **barometer** can be used to measure air pressure.	A **flow meter** can be used to measure river speed.	A **sound level meter** is used to measure noise pollution in different parts of a town.
A **thermometer** can be used to measure **temperature**.	A **metre stick** can be used to measure river depth.	You can measure the economic impact of an out-of-town shopping centre on shops located in the town centre **by comparing income**.
An **anemometer** can be used to measure wind speed.	A **measuring tape** and ranging poles can be used to measure width.	You can measure the width of footpaths for evidence of erosion, e.g. around school footpaths, using **a measuring tape**.

Questionnaires

Questionnaires are used to gather information from large numbers of people, to be analysed at a later date. This evidence can be either quantitative (using numerical data), such as the number of lorries recorded in a traffic survey, or qualitative (using non-numerical information), such as a person's opinion of how noisy a road is. Figure 1.8 shows an example of a questionnaire.

LAND USE CONFLICTS QUESTIONNAIRE

Date: Time: Location:

Task: Politely approach a person and say 'excuse me, I'm doing some work for my Geography course at school. Would you mind answering a few questions please?'

Ask them the questions below and record the answers in the table.

Remember to say 'thank you' when they have answered all your questions.

---

1. Do you live locally?
2. Do many tourists visit this area?
3. When is the most popular time for tourists?
4. What benefits have tourists brought to the area?
5. What problems do tourists cause here?
6. What has been done to minimise the impact of tourism on the area?

Question	Person 1	Person 2	Person 3	Person 4	Person 5	Person 6
1						
2						
3						
4						
5						
6						

Figure 1.8: *An example of a questionnaire*

Field-sketching and taking photographs

Field-sketching is a good way of capturing on paper the feature of your area of study that you are most interested in, e.g. a meander or a swallow hole. **Photography** can be a very useful information gathering technique. For example, you could take photographs at different points in a town and compare the different land use zones, or you could obtain old photographs and compare how land use changes over the course of a river or at different parts of a coastline.

Interviewing

An in-depth discussion about a geographical issue involves a structured **interview**, e.g. an interview with a farmer about different land use conflicts in and around his or her farm. Interviews can be filmed, sound recorded or logged in notes. Interviewing elderly people allows you to find out how things have changed over time as they will have lived through the changes, e.g. inner city redevelopments. Interviews with tourists enable you to find out their opinions of tourist facilities in an area.

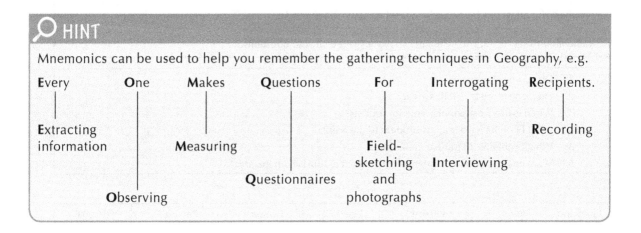

$\mathcal{O}$ HINT

Mnemonics can be used to help you remember the gathering techniques in Geography, e.g.

Every	One	Makes	Questions	For	Interrogating	Recipients.
Extracting information		Measuring		Field-sketching and photographs	Interviewing	Recording
	Observing		Questionnaires			

Methods of processing data collected in Geography

Graphs

Graphs are used to support evidence of your written work. They are a good way of displaying data that you have gathered in the field, on the internet or from books. Graphs should be coloured to enhance them and make the bars or segments stand out.

Bar graphs

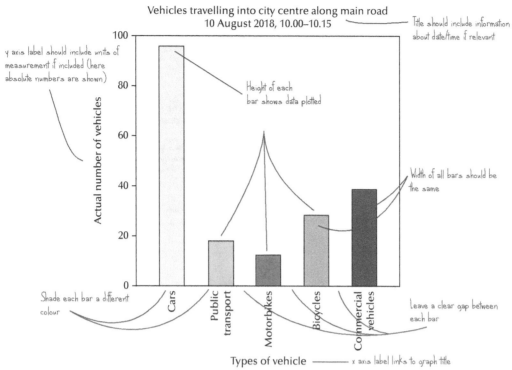

y axis label should include units of measurement if included (here absolute numbers are shown)

Title should include information about date/time if relevant

Height of each bar shows data plotted

Width of all bars should be the same

Shade each bar a different colour

Leave a clear gap between each bar

x axis label links to graph title

Figure 1.9: *How to create a bar graph*

Population pyramids

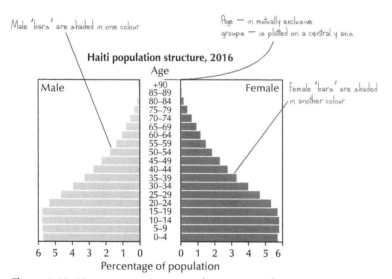

Male 'bars' are shaded in one colour

Age — in mutually exclusive groups — is plotted on a central y axis

Female 'bars' are shaded in another colour

Figure 1.10: *How to construct a population pyramid*

🔍 HINT

A **histogram** also has bars but there are no spaces between them. Histograms are used to show continuous data where there is a change over time, e.g. rainfall over a year.

🔍 HINT

A population pyramid is a double bar graph – you can see this if you turn it on its side.

15

Line graphs

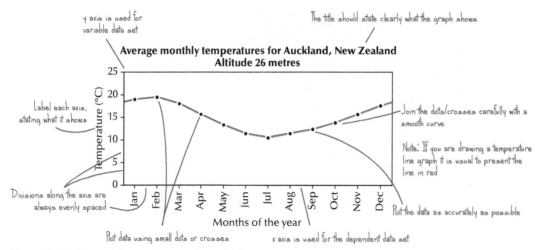

Figure 1.11: *How to draw a line graph*

<table>
<tr><td>

HINT

Use a pencil when drawing graphs so that any mistakes can be easily erased.

</td></tr>
</table>

<table>
<tr><td>

HINT

You can convert data by dividing each category by the total number and multiplying each answer by 360 to get the number of degrees for each category.

</td></tr>
</table>

<table>
<tr><td>

HINT

Remember to add a key or labels to show what each section represents.

</td></tr>
</table>

Pie graphs

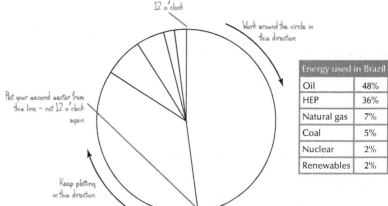

Energy used in Brazil	
Oil	48%
HEP	36%
Natural gas	7%
Coal	5%
Nuclear	2%
Renewables	2%

Types of energy used in Brazil

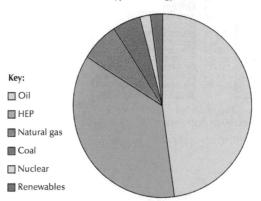

Key:
- ☐ Oil
- ▤ HEP
- ▨ Natural gas
- ▪ Coal
- ☐ Nuclear
- ▨ Renewables

Figure 1.12: *How to produce a pie graph*

Divided bar graphs

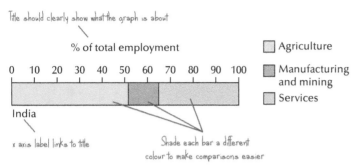

Title should clearly show what the graph is about

Shade each bar a different colour to make comparisons easier

x axis label links to title

Figure 1.13: *How to construct a divided bar graph*

Scatter graphs

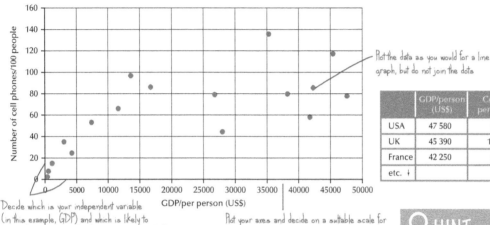

Plot the data as you would for a line graph, but do not join the dots

Decide which is your independent variable (in this example, GDP) and which is likely to be the dependent variable (cell phone ownership)

Plot your axes and decide on a suitable scale for each one. Note that the spaces between values must be the same along each axis

Figure 1.14: *How to construct a scatter graph*

Climate graphs

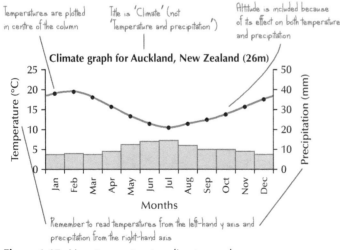

Temperatures are plotted in centre of the column

Title is 'Climate' (not 'Temperature and precipitation')

Altitude is included because of its effect on both temperature and precipitation

Remember to read temperatures from the left-hand y axis and precipitation from the right-hand axis

Figure 1.15: *How to construct a climate graph*

Classifying and tabulating

Information can be classified or sorted into groups, e.g. developed and developing countries. **Tabulating** involves drawing a table to show data or information, e.g. the advantages and disadvantages of tourism on the landscape. Tables are a good way of organising and summarising data.

Making maps

Maps are a means of showing links between features. They are a selective way of showing information and can reveal patterns, e.g. land use maps show how the land is used in an urban area. **Dot distribution maps** allow you to show population density as shown in Figure 1.16.

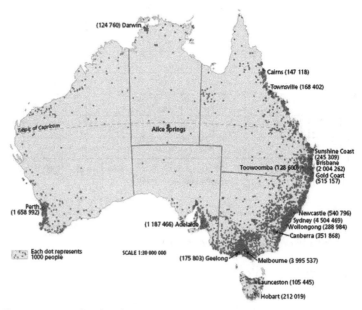

Figure 1.16: *A dot distribution map*

Annotating

To **annotate** means to label. You can label a photograph, field-sketch or map to highlight key points including relationships between old and new photographs, patterns on maps and processes on sketches.

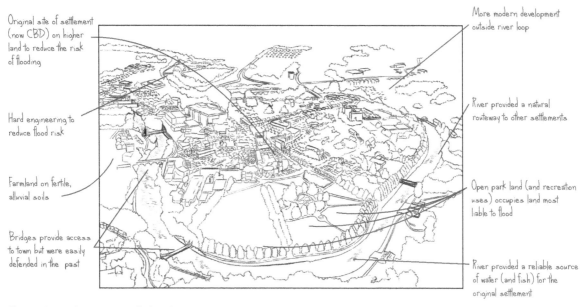

Original site of settlement (now CBD) on higher land to reduce the risk of flooding

Hard engineering to reduce flood risk

Farmland on fertile, alluvial soils

Bridges provide access to town but were easily defended in the past

More modern development outside river loop

River provided a natural routeway to other settlements

Open park land (and recreation uses) occupies land most liable to flood

River provided a reliable source of water (and fish) for the original settlement

Figure 1.17: *An annotated sketch*

Cross-sections and transects

A **cross-section** is a side view of the landscape. It shows changes in relief and slope between two points on an OS map. **Transect** diagrams are often used with cross-sections. When each part of a cross-section is annotated it becomes a **transect**. As cross-sections show the relief of the land, labels are often added to identify features like **geology**, soils, landforms or land uses of different parts of the cross-section.

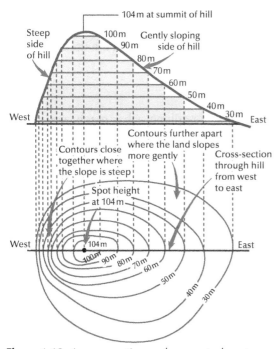

Figure 1.18: *A cross-section and connected contours*

Make the Link

You should be able to interpret a cross-section on an OS map.

🔍 HINT

Mnemonics can be used to help you remember the processing techniques in Geography, e.g.

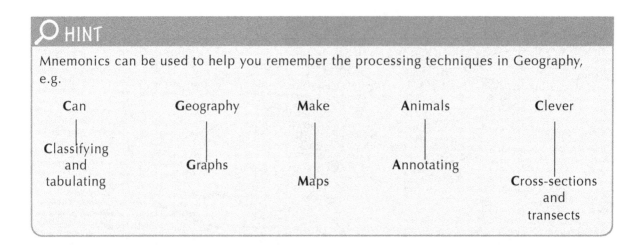

Can	Geography	Make	Animals	Clever
Classifying and tabulating	Graphs	Maps	Annotating	Cross-sections and transects

💥 Make the Link

In Maths and Science you will have learned how to draw graphs. These skills should help you to select appropriate titles, scales and axes for graphs.

🔍 HINT

You can develop your problem solving skills by looking at different sets of data and working out the most appropriate graph to draw.

🔍 HINT

When answering questions which ask you to describe graphs in detail, it is important to quote figures to support your answer.

💥 Make the Link

In Maths and Science you also learn how to use graphs and manipulate statistics.

🔍 HINT

Always read the title of the graph to help you work out what it is about.

🚦 Activity 2: Group activity

In groups, investigate the topics in the book where each of the gathering and processing techniques are applicable. Draw summary spider diagrams:

1. for gathering techniques
2. for processing techniques.

Extracting and interpreting geographical information

Graphs and tables help you to summarise data, draw conclusions and see patterns at a glance. You must be able to extract and interpret information from them. Different forms of graphs are used to present different types of data.

Describing bar graphs

Bar graphs show totals. When describing a bar graph, you should include:

- The highest and lowest amounts shown by the tallest and shortest bars.
- A comparison of the bars by noting differences between them.
- The total, e.g. total rainfall per month or per year.

Describing population pyramids

Population pyramids show the population structure of a country. When describing a population pyramid, you should include:

- The overall shape of the graph – bullet shaped, indicating a **developed country** or pyramid shaped, indicating a **developing country**.
- The width of the base of the pyramid, e.g. wide or narrow.

- The width of the top of the pyramid.
- The active population (16–64).
- The dependency ratio: the number of active people compared to under 15s and over 65s.
- Anomalies: are there any bars that are unusually shorter or longer than the overall trend?

HINT

Remember to identify the numbers and quote them – they may be percentages of the total population or actual numbers of people.

Describing line graphs

Line graphs can have one or more lines. When describing a line graph, you should include:

- The general pattern of the line showing an increase or decrease.
- A trend or change over time.
- The rate of change, highlighted by the steepness of the line.

HINT

Different colours can be used to differentiate each line on a **multiple line graph**.

Describing pie graphs

Pie graphs are commonly used to display percentages and show how an amount is shared out. When describing a pie graph, you should include:

- The highest and lowest shown by the biggest and smallest segment.
- The differences indicated by the size of each segment.
- The proportion or percentage of one segment compared to another.

HINT

Use the different colours to help you describe the size of each segment.

Describing divided bar graphs

Divided bar graphs are also used to display percentages and show proportions by the size of each bar. When describing a divided bar graph, you should include:

- The highest and lowest shown by the biggest and smallest bar.
- The differences indicated by the size of each bar.
- The proportion or percentage of one bar compared to another.

Describing scatter graphs

Scatter graphs are used to show a link between two sets of data. When describing scatter graphs, you should include:

- The connection between the two sets of data on the graph.
- Any pattern on the graph, shown by a best fit line.
- Positive or negative relationships.

Describing climate graphs

Climate graphs show temperature and rainfall for an area over one year. The rainfall bars are joined together like a histogram because the data is continuous. When describing a climate graph, you should include:

HINT

Read all labels carefully to help you work out what the graph is about.

⁂ Make the Link

If a question asks you to give reasons for the climate data, use your knowledge of the effects of climatic factors, e.g. relief, altitude, latitude and nearness to oceans (see Weather chapter).

- The highest and lowest temperature figures and the month they occurred.
- The temperature range (highest–lowest).
- The average yearly temperature (add up all the numbers and divide by 12).
- The highest and lowest rainfall figures and the month they occurred.
- The total annual rainfall (add up all the numbers for rainfall over the 12 months).
- The average rainfall (add up all the numbers and divide by 12).

🔍 HINT

You can develop your thinking skills by looking at a number of different types of graphs and describing the different trends they show.

GO! Activity 3: Paired activity

1. In pairs, research data and draw at least six different graphs. Graphs should be labelled and coloured.
2. Make an illustrated **mind map** to show the different types of graphs.
3. Your mind map should include:
   (a) An eye-catching title.
   (b) A list of bullet points beside each graph detailing the key features.

Research skills in Geography

While field work allows you to gather information first hand, researching secondary data for an investigation enables you to collect information from secondary sources such as books, the internet and TV programmes. It is important that this data is reliable and it is essential to check that the information that you have researched is correct. It is also necessary to keep a note of your sources to reference your information. Your local library or school library is a great place to start your research. When conducting an investigation, there are a number of stages that you need to work through. The investigation model shown in Figure 1.19 is a generic way to go about producing a report in Geography.

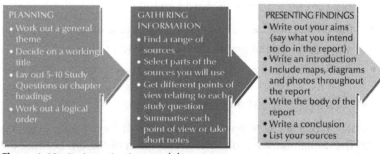

Figure 1.19: *An investigation model*

Planning the investigation

1. Decide on a topic and sections/chapters.
2. Make up an eye-catching title.
3. Decide on a focus: investigation questions should begin with where, when, what, who and why.
4. Decide how you are going to present your work, e.g. poster, PowerPoint or project.
5. Plan a layout, including the information, maps and pictures you will need.

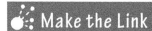

HINT

For specific and up-to-date advice on completing your N4 Added Value Unit or N5 Assignment, please refer to SQA's website.

Gathering information

1. Collect information from different sources such as textbooks, encyclopaedias, newspapers, DVDs, TV programmes, the internet and the news.
2. Make notes on the focal points of your investigation. You can highlight or underline key words or write/type a list using bullet points.
3. Notes must be written in your own words – you are not allowed to copy information word for word, unless you are quoting someone. You should read information, learn from it and then put what you have learned into your own words.
4. Cross-reference key facts and figures from different sources to ensure they are reliable.
5. Add value to diagrams, maps and photographs by labelling them to include extra information.

Make the Link

Gathering information for investigations helps to enhance your literacy skills, e.g. note-taking and spelling.

Presenting your findings

1. Arrange your information and diagrams into sections/chapters in a logical order.
2. Write aims and an introduction at the start.
3. Diagrams, maps and photographs should be clear with a title and labels. They should also be referred to within the text.
4. The middle sections/chapters should contain your key findings, maps and diagrams.
5. Write a conclusion at the end of your investigation to sum up the key points.
6. Include a reference list to identify your sources.

Make the Link

Investigations allow you to apply the skills that you have learned in different subjects such as Maths and English as you can put into practice your literacy and numeracy skills.

Summary

In this chapter you have learned:

- Seven techniques for gathering geographical information.

- Five methods of processing data collected in Geography.

- How to conduct research into various aspects of Geography and the stages that you need to work through when producing a report.

You should have developed your skills and be able to:

- Use mapping skills related to Ordnance Survey maps including: 4 and 6 figure grid references, measure distance using scale, interpret relief and contour patterns, use maps in association with photographs, field sketches, cross sections and transects.

- Extract, interpret and present numerical and graphical information which may be: graphs, tables, diagrams or maps.

End of chapter questions

Map skills

National 4 questions

(a) Why do geographers use a compass?

(b) What is a map scale used for?

(c) Why do OS maps have a grid system?

(d) What is the difference between using 4- and 6-figure grid references?

(e) What are the three ways of showing height on an OS map?

(f) What is a ridge?

(g) Draw the contour pattern of a valley.

(h) What is the symbol for a cliff on an OS map?

National 5 questions

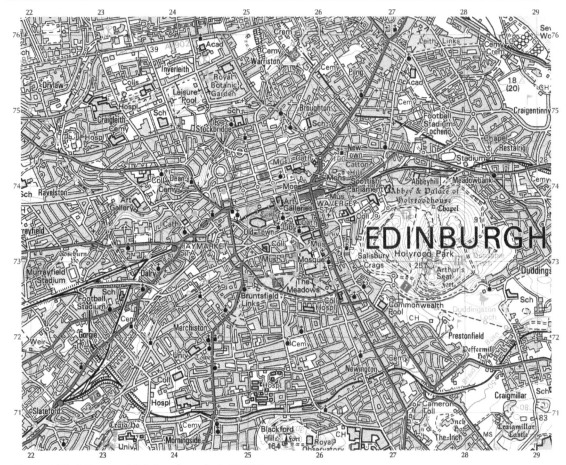

Figure 1.20: *OS map of Edinburgh, scale 1:50 000*

Study the OS map of Edinburgh and answer the following questions:

(a) Name the symbol marking the height of the land in grid square 2875.

(b) What is located in grid square 2475?

(c) Give a 4-figure grid reference for Murrayfield Stadium.

(d) Name the tourist attraction found at grid reference 257737.

(e) Give a 6-figure grid reference for the bus station in grid square 2574.

(f) Name the building located at grid reference 288726.

(g) Give a 6-figure grid reference for the train station in grid square 2271.

National 5 exam-style questions

You can find sample answers to these exam-style questions on the Leckie website:
https://collins.co.uk/pages/scottish-curriculum-free-resources

1   Study the OS map of Edinburgh

Give **map evidence** to prove that part of Edinburgh's CBD is located in grid square 2573.

(4 marks)

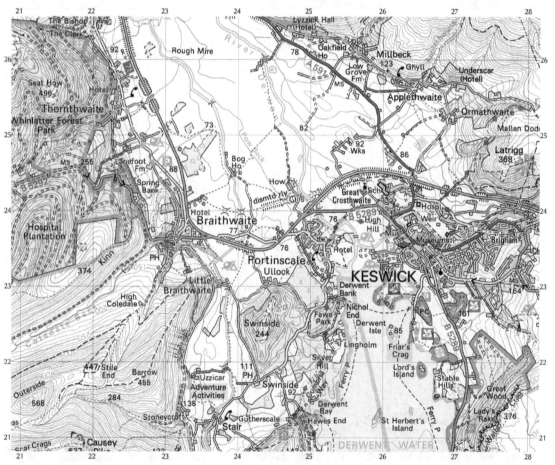

Figure 1.21: *OS map of Keswick, scale 1:50 000*

2 Study the OS map of Keswick.

Discuss the land use conflicts in and around Keswick.

(6 marks)

3 Study the OS map of Keswick.

Describe the different tourist attractions on the map.

(5 marks)

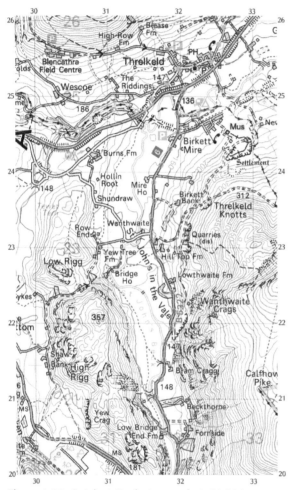

Figure 1.22: *St John's Beck river, scale 1:50 000*

4 Study the OS map of St John's Beck river.

Describe the St John's Beck (river) and its valley from 319200 to 329250.

(5 marks)

Gathering and processing techniques

National 4 questions

(a) List **eight** different ways of gathering information in Geography.

(b) How would you conduct a traffic survey?

(c) Name **three** aspects of an environmental quality survey.

(d) List **three** things you can measure in the field in Geography.

(e) What is the difference between an interview and a questionnaire?

(f) List **five** different ways of processing information in Geography.

(g) Name **six** different types of graphs used in Geography.

(h) Why is it important to label graphs that you have drawn?

(i) What is the difference between a bar graph and a histogram?

(j) What **two** things does a population pyramid show?

National 5 questions

(a) Draw a bar graph using the information in the table below:

COUNTRY	PERCENTAGE OF PEOPLE EMPLOYED IN AGRICULTURE
Brazil	20·0
China	38·1
India	52·0
Nigeria	70·0
UK	1·4
USA	0·7

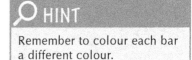

HINT

Remember to colour each bar a different colour.

(b) Describe what the bar graph you have drawn shows.

(c) Draw a line graph using the information in the table below:

YEAR	WORLD POPULATION CHANGE (BILLIONS)
1960	3·0
1970	3·75
1980	4·50
1990	5·25
2000	6·0
2010	6·95
2020 (FORECAST)	7·50
2030 (FORECAST)	8·25
2040 (FORECAST)	8·95
2050 (FORECAST)	9·25

> ### 🔍 HINT
> Don't forget to add a title to your graph.

(d) Describe what the line graph you have drawn shows.

(e) Draw a pie graph using the information in the table below:

DIFFERENT RENEWABLE ENERGIES AS A PERCENTAGE OF THE GLOBAL TOTAL	PERCENTAGE	DEGREES ON PIE GRAPH
Hydro	63	227
Geothermal	3	11
Wind	5	18
Solar	7	25
Power from plants (including biogas)	22	79

> ### 🔍 HINT
> Remember to add labels to each segment.

(f) Describe what the pie graph you have drawn shows.

(g) Draw a scatter graph using the information in the table below:

COUNTRY	Gross Domestic Product (average per person)	Adult Literacy Rate (percentage)
Bangladesh	1770	48
Brazil	11000	89
Canada	40000	99
Ethiopia	1000	43
France	33000	99
Nigeria	2000	68
Pakistan	2000	49
UK	35000	99

HINT

Don't forget to label both the X and Y axis.

(h) Describe what the scatter graph you have drawn shows.

(i) Draw a climate graph using the information in the table below:

MONTH	Jan	Feb	Mar	Apr	May	June	July	Aug	Sept	Oct	Nov	Dec
RAINFALL (mm)	235	225	240	200	180	100	125	130	150	160	160	180
TEMPERATURE (°C)	30	20	28	28	29	29	30	30	30	30	29	29

(j) Describe what the climate graph you have drawn shows.

HINT

Use graph paper to help you draw more accurate graphs.

NATIONAL 5 EXAM-STYLE QUESTIONS

You can find sample answers to these exam-style questions on the Leckie website:
https://collins.co.uk/pages/scottish-curriculum-free-resources

1

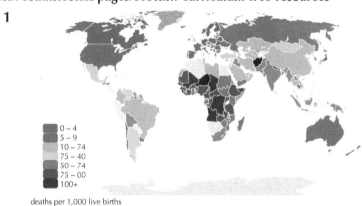

deaths per 1,000 live births

Diagram Q1: *Global infant mortality rates per thousand live births*

Study Diagram Q1 above.

Describe, in detail, global infant mortality rates.

(4 marks)

2

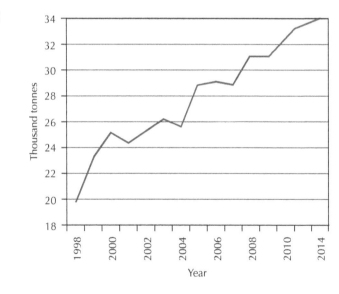

Diagram Q2: Rice production in Bangladesh, 1998–2014

Study Diagram Q2 above.

Describe the changes in rice production in Bangladesh.

(4 marks)

3

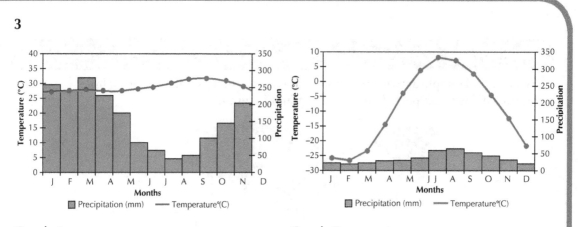

Graph A   Graph B

Diagram Q3A: *Climate graphs for two selected regions*

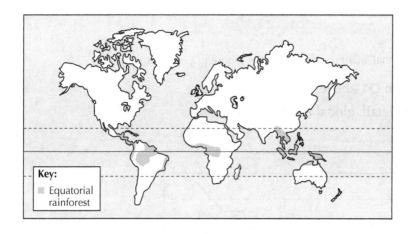

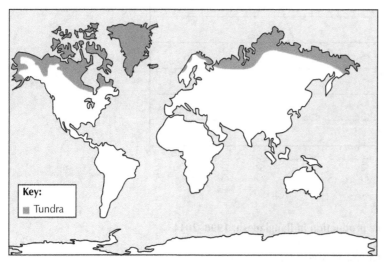

Diagram Q3B: *Location maps of two selected climate regions*

Study Diagrams Q3A and Q3B above.

Describe **and** explain the differences between the climate graphs.

(6 marks)

4 **Diagram Q4:** Gross Domestic Product (GDP) and adult literacy for selected countries

COUNTRY	GDP	ADULT LITERACY
Canada	40000	99
Pakistan	2000	49
France	33000	99
South Africa	11000	86
Ethiopia	1000	43
UK	35000	99

Study Diagram Q4 above

Use the information in Diagram Q4 to **complete** the scatter graph below.

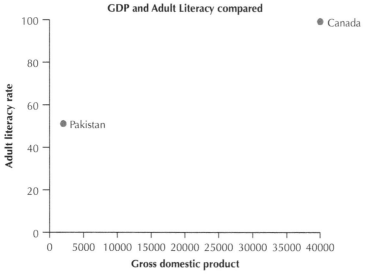

Diagram Q4 scatter graph

(4 marks)

LEARNING CHECKLIST

Now that you have finished the **Geographical skills** chapter, complete a self-evaluation of your knowledge and skills to assess what you have understood. Use traffic lights to help you make up a revision plan to help you improve in the areas you identified as red or amber.

- Use a compass to give different directions on an OS map.

- Measure distance on a map and use a scale line to calculate the real distance on the ground.

- Give 4-figure grid references to locate a grid square on an OS map.

- Give 6-figure grid references to pinpoint the location of a symbol on an OS map.

- Identify different ways of showing relief on an OS map including:

  » spot heights

  » trig points

  » contour lines

- Identify different landforms according to their contour patterns on an OS map.

- List seven gathering techniques used in Geography.

- List five processing techniques.

- Draw the following graphs:

  » bar

  » population pyramid

  » line (single or multiple)

  » pie

  » divided bar

» scatter

» climate

- Extract and interpret the key things that each graph shows:

  » bar

  » line

  » pie

  » scatter

  » climate

- Extract and interpret information from topographical maps, e.g. dot distribution maps.

- Undertake a geographical investigation by:

  » planning

  » gathering

  » presenting

Glossary

Map skills

Cliff: A very steep (vertical) slope.

Compass: An instrument used to measure direction.

Confluence: The meeting point of two rivers.

Contour lines: Thin orange/brown lines on an OS map used to show the height of the land at 10-metre intervals.

Contour patterns: The shape of the land outlined in brown/orange lines on an OS map.

Eastings: Lines running N/S on an OS map.

Escarpment: A ridge with one side that is much steeper than the other.

Estuary: The area where river water meets sea water when a river enters the sea.

Flood plain: The flat land surrounding a river in its lower course.

Grid references: The number of a grid square (4-figure grid reference) or a specific symbol (6-figure grid reference) on an OS map.

Knoll: A small hill.

Landforms: The shape of the physical landscape.

Meander: A bend in a river.

Northings: Lines running W/E on an OS map.

Ordnance Survey (OS) Map: A specific type of map produced by a company called Ordnance Survey.

Pass: A long valley providing a natural route way through a mountainous area.

Plateau: A flat-topped area of high land.

Ridge: A long, narrow area of high land.

Scale: The ratio located on a map that is used to help calculate the real distance between different places on the ground.

Spot height: A black dot with a number beside it marking the height of the land on an OS map.

Spur: A finger of high land that sticks out into an area of lower land.

Topographical map: A type of map showing cultural and natural features on the ground.

Triangulation pillar: A symbol on an OS map used to show the highest point of the land (shown by a concrete pillar on the ground). (Also called trig point.)

Valley: A narrow area of land that has higher land on both sides.

Gathering and processing techniques

Annotate: To label, e.g. a sketch of a physical feature.

Bar graph: A graph with different bars used for comparing different amounts.

Classify: To sort information into different groups or categories.

Climate graph: One graph containing a bar graph showing yearly precipitation and a line graph displaying yearly temperatures of a place.

Cross-section: A side-on view of the landscape.

Divided bar graph: One bar split into smaller sections to show percentages.

Dot distribution map: A map showing population density (in dots) in different areas, e.g. of a country.

Environmental quality survey: When different aspects of the environment are observed and recorded on a scale of 0–10, e.g. litter.

Field-sketch: To draw a feature of the landscape.

Gathering technique: A way of collecting geographical information, e.g. a questionnaire or field-sketch.

Geology: The study of different rocks.

Histogram: A bar graph used to show continuous data so there are no spaces between the bars.

Interview: An in-depth conversation with a person to extract information from them.

Line graph: A graph with a line, often used to show a trend over time.

Multiple line graph: A line graph that has more than one line.

Pie graph: A graph with different segments used to show how an amount is shared out.

Population pyramid: A double bar graph showing the age-sex composition of a country.

Processing technique: A way of representing geographical information, e.g. a bar graph or a labelled photograph.

Questionnaire: A written survey issued to people to fill in their answers to different questions.

Scatter graph: A graph with two sets of data plotted to establish a connection.

Tabulate: To draw a table.

Traffic survey: When vehicles are observed and recorded in a certain place over a time period.

Transect: When land use is shown on a cross-section.

Geography is the study of the people and landscapes, especially the impact of people and their activities on the environment. It is essential to understand the weather and its effects because it impacts on both the physical landscape such as the type of plants that grow in a particular area and the human landscape e.g. how farmers use their land.

It is important to study physical environments as the shape of the land and its landforms impact upon the way people interact with and make use of the natural environment. Physical landscapes include:

✤ Rivers and valleys

✤ Upland limestone

✤ Glaciated uplands

✤ Coastal landscapes

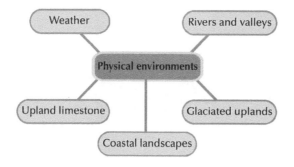

➤ These landscapes have a huge influence on different land uses including: Farming

➤ Forestry

➤ Industry

➤ Recreation and tourism

➤ Water storage and supply

➤ Renewable energy.

Think about how the weather affects you and the different landscape types that you have visited.

The Physical
Environments

2 Weather

Introduction

The **weather** is the current condition of the air around us. It is made up of different elements:

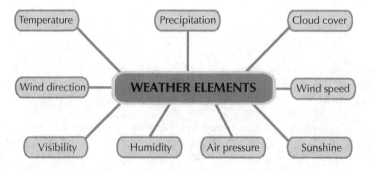

The idea of 'typical weather' is known as **climate** – the usual weather that a place experiences over an average period of 30–40 years. By averaging out elements of the weather such as **temperature** and **rainfall**, it is possible to identify what the weather is usually like in different places without being influenced by unusual weather events. Both weather and climate vary from place to place.

> **HINT**
>
> The climate of the rainforest is hot and wet as it experiences high temperatures and high rainfall throughout the year.

The effect of latitude, relief, aspect and distance from the sea on local weather conditions

Look at Figure 2.1, which shows the regional differences in summer and winter in the UK.

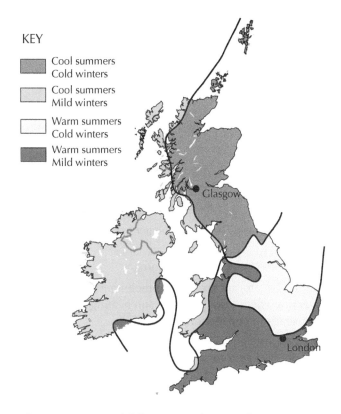

KEY

Cool summers
Cold winters

Cool summers
Mild winters

Warm summers
Cold winters

Warm summers
Mild winters

Glasgow

London

Figure 2.1: *Seasonal differences in the UK's climate*

The weather in the UK is affected by different factors.

Latitude

Latitude refers to the location of a place in relation to the **Equator**. It is measured in degrees. The value of the Equator is 0 degrees. Places nearer the **North Pole** (90°N) and South Pole (90°S) are typically colder and drier – they have high latitudes. Places that are closer to the Equator (0°) are usually warmer and wetter – they have low latitudes.

The reasons for the differences in temperature are:

1. The dark leaves of the trees in the equatorial rainforests located along the Equator absorb the sun's heat, making the temperature warmer. On the other hand, the snow and ice found at the poles reflect the sun's heat, making it colder.

2. The sun's rays have a thinner atmosphere to pass through at the Equator. This means there is less heat absorbed by the dust and gases in the atmosphere, making the temperature higher. The opposite is true at the poles.

3. Look at Figure 2.2. It shows that the sun's rays shine directly on the Equator, so heat is focused in that area, making it warmer. The curve of the Earth in the northern and southern hemispheres means that the sun's rays are more spread out at the poles, making it colder there.

Therefore in the UK, places which are closer to the North Pole e.g. Wick in northern Scotland are colder than places which are closer to the Equator e.g. Bournemouth in southern England.

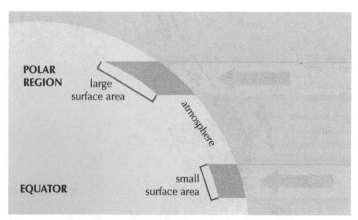

Figure 2.2: *The effect of latitude on weather conditions*

In the Natural Regions chapter, you will study convectional rainfall in more detail.

The reasons for the differences in **precipitation** are:

1. Places near to the Equator receive more precipitation (rain) due to low pressure zones which result from the sun heating the air at the Equator, which leads to **convectional rainfall**.

2. Places near to the North and South Poles receive less precipitation (usually snow) due to high pressure zones which result from cold air sinking. In the UK levels of precipitation experienced in different locations is complex due to a combination of factors including relief and distance from the sea.

Relief

This refers to the height and shape of the landscape. Places situated on flat, low land are warmer as temperatures decrease on average by 10°C for every 1000 metres in height. This is because there are fewer solid particles in the upper air to retain heat, so heat is lost to space. Places higher up are therefore colder and wetter. In the west of the UK the land is higher, creating more rainfall in the west.

Look at Figure 2.3. It shows how precipitation occurs over high land.

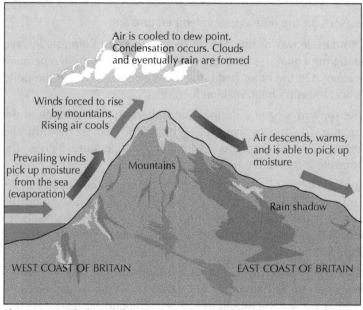

Figure 2.3: *Relief rainfall*

The reasons for the differences in rainfall in the UK are:

1. The **prevailing winds** are south westerly. These winds pick up moisture from the Atlantic Ocean.

2. Air is forced to rise over the high land on the west of the UK.

3. As the air rises it cools, condenses and water droplets form.

4. As more condensation occurs, the water droplets in the clouds get bigger and bigger.

5. Clouds burst and rain falls.

6. Air sinks and warms where the height of the land decreases, producing drier weather conditions on the east coast.

Aspect

This refers to the direction that a place lies in relation to the sun. Places that are south facing are usually warmer and sunnier because they face towards the Equator. They are also sheltered from cold northerly winds. Places which are north facing are colder because they are in the shadow of the sun and exposed to cold northerly winds from the North Pole.

Distance from the sea

This refers to the location of places either inland or along the coast. Places closer to the sea experience mild and wet weather conditions because clouds lose their moisture as they are blown across land. This helps to explain why the west coast of Scotland is much wetter than the east! Places further from the sea are drier and warmer in summer because the sun heats land areas quicker than seas and oceans. In winter time, temperatures are more extreme because the land cannot keep the heat. However, the seas and oceans can retain the heat in winter and this has a warming effect on coastal areas. The North Atlantic Drift is a warm ocean current that keeps the UK's weather warmer but wetter than it should be for its latitude.

The characteristics of the five main air masses affecting the UK

The weather in the UK is variable and hard for forecasters to predict. The changeable weather is attributed to the air masses which bring different weather conditions on a daily basis. An **air mass** is a large amount of air that moves from one area to another. The weather an air mass brings is associated with the region it has come from and the kind of surface it has travelled over.

There are five main air masses that affect the United Kingdom. Look at Figure 2.4 below.

Make the Link

In the Rivers section, you will study different types of farming throughout the course of a river. Think about how the weather can influence farming activities in different parts of the UK, e.g. high land and low land.

Make the Link

In the Global Issues section, you will study the impact of human activity on the equatorial rainforest and tundra environments. Think about how latitude, relief, aspect and distance from the sea affect the climate in those areas.

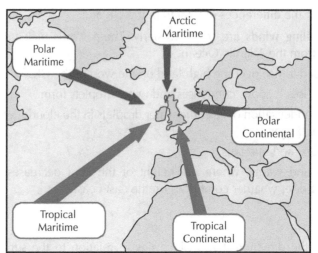

Figure 2.4: *Air masses*

> 🔍 HINT
>
> Use the coloured arrows to help you remember which air masses are warm and which ones are cold.

Figure 2.5 summarises the weather characteristics which an air mass brings.

Area air mass has come from:	Weather characteristics:
Arctic	Cold
Tropics	Warm
Land	Dry
Water	Wet

Figure 2.5: *The weather characteristics associated with air masses*

Each air mass brings specific weather conditions to the UK at different times of the year. (See Figure 2.6 below.)

Air mass	Where it has come from	Winter weather conditions	Summer weather conditions.
Arctic Maritime	Arctic Ocean	Very cold and snow	Cold and wet
Polar Continental	Northern Europe and Siberia in Russia	Very cold and wet (snow)	Hot, dry and sunny
Polar Maritime	North Atlantic Ocean	Cool, cloudy and rain showers	Cool, cloudy and rain showers
Tropical Continental	Northern Africa and Southern Europe	Mild, dry and sunny	Very hot, dry and sunny
Tropical Maritime	South Atlantic Ocean	Mild and wet	Warm and wet

Figure 2.6: *The seasonal weather associated with each air mass affecting the UK*

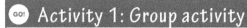

Activity 1: Group activity

Make up a **group poster** detailing how we get our weather in the UK.

You should include:

- An introduction describing what the poster is about.
- A map of the UK with the five air masses drawn on it.
- Written information and diagrams which show the effects of latitude, relief, aspect and distance from the sea on local weather conditions.
- Research on your own town or village and an explanation of weather conditions in your area.
- A table showing the different weather conditions the UK receives from different air masses in the summer and winter.
- A conclusion/summary to highlight the key points on your poster.

The characteristics of weather associated with depressions and anticyclones

Synoptic charts

A **synoptic chart** is a detailed weather map showing air pressure and different weather systems moving over the UK.

The main **weather element** shown on a synoptic chart is **air pressure**. Black lines called **isobars** join up areas of the same air pressure. The pattern these lines make shows areas of low pressure or high pressure. Tightly packed isobars symbolise low air pressure and high **wind speeds**. Spaced out isobars signal high air pressure and light winds/calm conditions. Fronts can also be found on synoptic charts and form part of a low pressure system called a **depression**. (See Figure 2.7.)

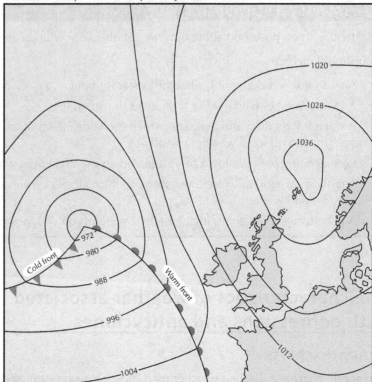

WEATHER MAP, NW EUROPE, 15TH JANUARY 2016

Figure 2.7: *A synoptic chart for the UK*

HINT

The date gives a clue to the likely weather conditions.

Make the Link

Handling information is a common task in many subjects including Maths and Physics. The basic skills are the same but the contexts are different.

Weather station circles

Synoptic charts often have symbols on them which describe the weather conditions at different places. Examine the example of the weather station circle in Figure 2.8 and the key (Figure 2.9):

HINT

Remember the 8 point campass to help you work out wind direction.

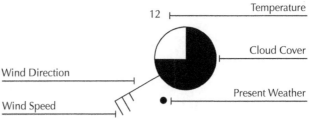

Figure 2.8: *A weather station circle*

Symbols used on a weather chart

Symbol	Precipitation	Circles	Cloud Cover	Circles	Wind Speed
🌢	Drizzle	○	Clear Sky	◎	Calm
▽	Shower	◐	One Okta	○—	1-2 Knots
●	Rain	◕	Two Oktas	○—┐	5 Knots
✶	Snow	◑	Three Oktas	○—└	10 Knots
△	Hail	◑	Four Oktas	○—┐	15 Knots
⦓	Thunderstrom	◑	Five Oktas	○—┐	20 Knots
🌢▽	Rain shower	◕	Six Oktas	○—▶	50 Knots or more
⁝●	Heavy Rain	◖	Seven Oktas		
●✶	Sleet	●	Eight Oktas		
✶▽	Snow Shower	⊗	Sky Obscured		
≡	Mist				
≣	Fog				

Figure 2.9: *Key for weather station symbols*

Low pressure systems – depressions

Our weather is dominated by the passage of depressions across the UK – about 100 depressions pass over in one year. A **depression** is an area of **low** air pressure which moves in an **easterly** direction across the UK. Depressions bring cloudy, windy and rainy conditions. They are formed over the Atlantic Ocean when one air mass meets another. The two air masses do not usually mix due to the differences in their temperature and density. The boundary between two air masses is called a **front**.

Low pressure systems can be identified from a synoptic chart by:

- Isobars showing air pressure falling towards the centre from about 1004mb
- Isobars which are close together
- Weather fronts – both warm and cold
- Likely occluded fronts.

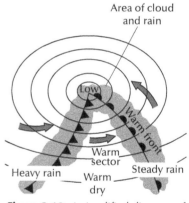

Figure 2.10: *A simplified diagram of a depression*

🔍 HINT

Fronts are bands of clouds which usually bring rain.

Depressions – weather fronts

Warm front

A depression is made up of a **warm front** and a **cold front**. The warm front is the first to pass over the UK. A warm front happens when a warm air mass meets a cold air mass. The warm air rises above the cold air forming **nimbostratus** clouds. Warm fronts are essentially a band of clouds which bring steady, continuous rain. Figure 2.10 shows the formation of a warm front.

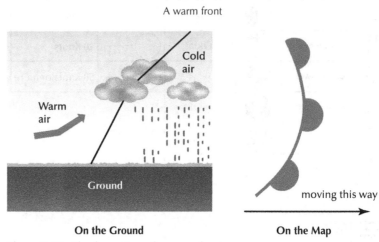

Figure 2.11: *The formation of a warm front*

Cold front

The next front to pass over is the cold front. This occurs when cold air meets warm air and the cold air pushes the warm air upwards. The warm air is forced to cool and condense quickly, forming large, towering **cumulonimbus** clouds. Cold fronts bring heavy rain showers. Figure 2.12 shows the formation of a cold front:

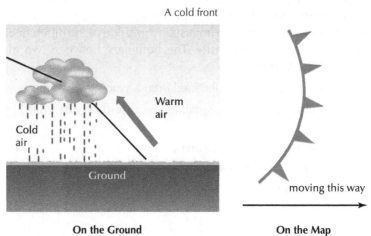

Figure 2.12: *The formation of a cold front*

Occluded front

When the cold front catches up with the warm front the result is an **occluded front**. The air mixes and causes weather conditions to be very unsettled. Occluded fronts bring sudden downpours of heavy rain and very strong winds. Figure 2.12 shows the appearance of an occluded front on a synoptic chart.

Depressions – features

- **Isobars**: Tightly packed isobars indicate strong winds. This is due to a rapid change in air pressure.

- **Wind:** Winds blow anti-clockwise and along the isobars. **Wind direction** can be worked out by following the isobars in an anti-clockwise direction.

- **Rain:** Warm fronts bring steady continuous rainfall as the warm air slowly rises up over the colder air at the warm front. Cold fronts bring heavy rain as the warm air is undercut by the cold air and forced up rapidly at a cold front.

- **Temperature:** In general, the **warm sector** brings warmer temperatures and the **cold sector** brings colder temperatures.

Figure 2.13 shows a basic version of a depression. The labels outline the weather linked to each part of the depression.

Occluded front

moving this way

On the Map
Figure 2.13: *The appearance of an occluded front on a synoptic chart*

High pressure systems – anticyclones

An **anticyclone** is an area of high pressure which brings long spells of fine, dry and settled weather. Clouds do not form as the cold air is sinking in a high air pressure system.

High pressure systems can be identified from a synoptic chart by:

- Isobars showing pressure increasing outwards from the centre and above 1008mb

- Widely spaced isobars

- No fronts/clouds

- No rain due to lack of clouds.

Anticyclones – features

- **Isobars** are widely spaced indicating gentle winds. Few isobars indicate calm conditions. This is due to a very gradual change in air pressure. These systems can remain in place for several days.

- **Wind.** Winds blow clockwise in high air pressure systems and along the isobars. You can work out the wind direction by following the isobars in a clockwise direction. Wind blows gently when isobars are widely spaced.

- **Dry conditions (no rain).** Skies are clear with very little cloud so no rain falls because cold air sinks in a high pressure system/anticyclone.

- **Temperature.** In summer anticyclones, temperatures are high and heat-wave conditions can occur. In winter anticyclones, temperatures are cold and there is a high chance of frost and occasional mist and fog.

Look at Figure 2.15. It shows what an anticyclone looks like on a synoptic chart.

HINT

Notice the differences between figure 2.14 and figure 2.15 e.g. no fronts in an anticyclone.

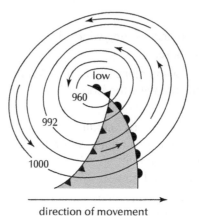

direction of movement

Figure 2.14: *A view of what a depression looks like on a synoptic chart*

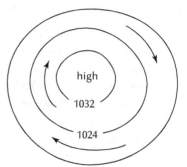

Figure 2.15: *A view of what an anticyclone looks like on a synoptic chart*

GO! Activity 2

Present your own weather forecast using a **poster or PowerPoint** presentation.
You should include:

- A synoptic chart of the UK complete with isobars, a depression **or** anticyclone and weather station circles.
- Music/sound.
- A theme, e.g. summer or winter.

🔵 Activity 3 (National 5)

Make an individual PowerPoint presentation showing how the weather is affected by depressions in your local area. Compare the weather conditions with another place in Europe that is experiencing an anticyclone.

You must include:

- A synoptic chart of Europe with a depression over the UK and an anticyclone over other parts of Europe.
- A date at the top of the chart – this will affect what you say about temperatures.
- The location of your town or village and another place in Europe (mark with a black dot and write the names of the places beside them).
- A weather station circle showing the weather conditions for your town or village.
- A detailed description of the weather station circle.
- An explanation of why the weather conditions are being experienced.
- The location of a place in Europe with an anticyclone.
- A weather station circle showing the weather conditions for your chosen place in Europe, e.g. Paris in France or Barcelona in Spain.
- A detailed description and explanation of the weather conditions being experienced.
- A comparison between both sets of weather conditions in your two areas.

Summary

In this chapter you have learned:

- The nine weather elements.
- What effect latitude, relief, aspect and distance from the sea have on our weather.
- The names and characteristics of the five main air masses affecting the UK.
- The characteristics of weather associated with depressions and anticyclones.

You should have developed your skills and be able to:

- Draw and label weather station circles.
- Interpret synoptic charts.

End of chapter questions

National 4 questions

(a) Match the following heads and tails and write the correct answers in sentences:

HEADS	TAILS
Latitude	Places closer to the sea receive more rainfall.
Relief	Places which are north facing are colder and shadier.
Aspect	Places which are higher up are colder and wetter.
Distance from sea	Places closer to the Equator are warmer.

(b) What makes mountainous areas cold and wet?

(c) Why are south facing places warmer and sunnier than north facing places?

(d) What type of weather does the Tropical Maritime air mass bring in summer?

(e) What type of weather does the Polar Continental air mass bring in winter?

(f) Look at the diagram below and answer the following questions:

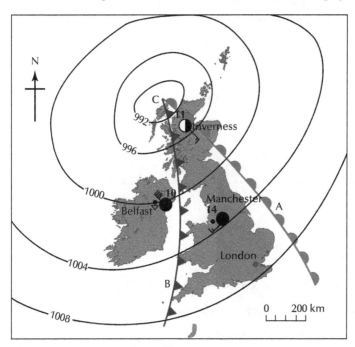

Diagram Q1: *Synoptic chart of the UK*

(i) Name the fronts labelled A and B.

(ii) Is Area C an area of high pressure **OR** low pressure?

     (iii) What is the air pressure, temperature, cloud cover, precipitation (rainfall), wind speed and wind direction at the following places:

- Inverness
- Manchester
- Belfast

     (iv) Explain what will happen to London's weather in the next 6 hours.

(g) Draw a weather station circle for the following weather conditions:

     (i) Cloud cover = 2 oktas

     (ii) Wind direction = SW

     (iii) Wind speed = 5 knots

     (iv) Temperature = 15°C

(h) Draw and colour the synoptic chart symbols for:

     (i) Warm front

     (ii) Cold front

     (iii) Occluded front

(i) In a depression, what type of weather conditions are brought by:

     (i) Warm fronts

     (ii) Cold fronts

     (iii) Tightly packed isobars

(j) Describe the weather conditions brought by an anticyclone in:

     (i) Winter

     (ii) Summer

National 5 questions

(a) Give reasons for the differences in weather conditions on the west and east coast of the UK.

(b) Describe how latitude affects temperature **and** rainfall.

(c) Name the five main air masses affecting the weather in the UK and describe the weather conditions brought by each air mass in both summer and winter.

(d) Give the **advantages** and **disadvantages** of a Tropical Continental air mass over the UK in summertime.

(e) Outline the type of weather brought by an Arctic Maritime air mass over the UK in wintertime and suggest the type of problems that this could cause for people.

(f) Describe how an occluded front is formed.

(g) **Explain** how weather changes with the passage of a depression. You should refer to temperature, rainfall, wind speed and air pressure in your answer.

(h) **Describe** the similarities and differences between an anticyclone in the summer and winter.

National 5 exam-style questions

You can find sample answers to these exam-style questions on the Leckie website:
https://collins.co.uk/pages/scottish-curriculum-free-resources

1

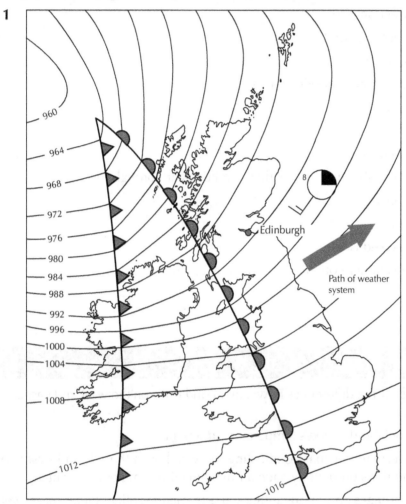

Diagram Q1: *a synoptic chart of the UK, April 2016*

Study Diagram Q1 above.

Describe and **explain** how the weather conditions will change in Edinburgh over the next 24 hours.

(5 marks)

2

1200 hours 10th November

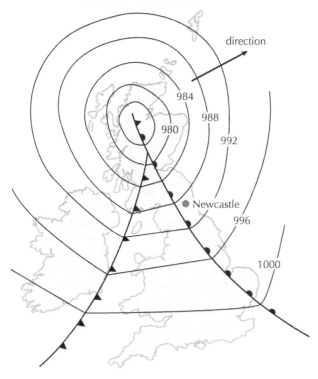

direction

984

988

980

992

● Newcastle

996

1000

⟶ Path of depression

Diagram Q2: *a depression over the UK.*

Study Diagram Q2 above:

Explain in detail what will happen to the weather in Newcastle in the next twenty-four hours. You should make reference to temperature, wind speed and direction, precipitation and air pressure.

(5 marks)

3 a) 22

b) 6

c) 14

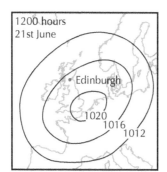

1200 hours
21st June

● Edinburgh

1020
1016
1012

Diagram Q3: *an anticyclone over NW Europe.*

Study Diagram Q3 above.

Which of the three weather station plots shows the weather conditions most likely experienced in Edinburgh at noon on 21 June?

Explain your choice in detail.

(5 marks)

LEARNING CHECKLIST

Now that you have finished the **Weather** chapter, complete a self-evaluation of your knowledge and skills to assess what you have understood. Use traffic lights to help you make up a revision plan to help you improve in the areas you identified as red or amber.

- Name the nine weather elements.

- Describe and explain the effect of latitude on local weather conditions.

- Describe and explain the effect of relief on local weather conditions.

- Describe and explain the effect of aspect on local weather conditions.

- Describe and explain the effect of distance from the sea on local weather conditions.

- Name the five main air masses which affect the weather in the UK.

- List the characteristics of the five main air masses affecting the UK.

- Describe the differences between the winter and the summer weather characteristics for each of the five air masses.

- Discuss what is shown on a synoptic chart.

- Draw different weather station circles highlighting different weather conditions.

- List the information shown on weather station circles.

- Describe the characteristics of weather associated with depressions.

- Explain the weather associated with depressions.

- Describe the characteristics of weather associated with anticyclones.

- Explain the weather associated with anticyclones.

Glossary

Air mass: A large volume of air which travels from one area to another.

Air pressure: The weight of the air in the atmosphere.

Anticyclone: An area of high air pressure which can last for several days and brings fine and settled weather.

Arctic Circle: An imaginary line of latitude that circles the globe at latitude 66 degrees north of the Equator.

Arctic Maritime: A very cold and wet air mass.

Aspect: The location a place lies in relation to the sun.

Climate: The usual weather that a place experiences over an average period of 30–40 years.

Cloud cover: The fraction of the sky (measured in eighths) that has cloud.

Cold front: A band of cloud formed when warm air meets cold air, and the cold air pushes the warm air upwards.

Cold sector: An area of colder air that surrounds the wedge of warm air in the warm sector of a depression.

Convectional rainfall: When the sun heats the air, warm, moist air rises, cools, condenses and forms clouds. These large, towering clouds burst and heavy rain falls.

Cumulonimbus: A type of cloud which is thick and dark and brings heavy rainfall.

Depression: An area of low pressure which moves from west to east in the northern hemisphere and alters weather conditions as it moves over an area.

Equator: An imaginary line of latitude that divides the Earth into a northern hemisphere and a southern hemisphere.

Humidity: The amount of water vapour in the air, measured in percentages.

Isobars: Lines on a synoptic chart which join places of equal air pressure.

Latitude: A type of coordinate used by geographers to indicate the position a place lies north or south of the Equator on the Earth's surface.

Nimbostratus: A type of cloud that is dark, spread-out and low-lying.

North Pole: The place that is located at latitude 90 degrees north of the Equator. It is the northernmost point on the Earth's surface.

Occluded front: A band of thick cloud formed when the cold front catches up with the warm front.

Polar Continental: A cold and dry air mass.

Polar Maritime: A cold and wet air mass.

Precipitation: Any moisture that falls from the sky: rain, hail, sleet or snow.

Prevailing wind: The most common direction the wind blows from.

Rainfall: Water droplets that fall from clouds.

Relief: The height of the land above sea level, usually expressed in metres.

Relief rainfall: When moist air from the sea rises over a mountain, it gets cold and forms clouds which burst and rain falls.

Sunshine amount: The number of hours the sun shines in one day.

Synoptic chart: A detailed weather map containing different lines and symbols.

Temperature: How hot or cold the air is outside.

Tropical Continental: A warm and dry air mass.

Tropical Maritime: A warm and wet air mass.

Tropics: A region of the Earth surrounding the Equator. The Tropic of Cancer is located at latitude 23·5 degrees north of the Equator. The Tropic of Capricorn is located at latitude 23·5 degrees south of the Equator.

Visibility: The distance at which an object can be seen, usually measured in metres.

Warm front: A band of cloud formed when warm air meets cold air, and the warm air rises above the cold air.

Warm sector: An area of warm air between the warm front and the cold front in a depression.

Weather: The current condition of the air.

Weather station circle: A symbol on a synoptic chart which shows different weather elements such as temperature and wind speed.

Weather element: An aspect of the condition of the air, e.g. temperature or rainfall.

Wind direction: The direction from which the wind is blowing.

Wind speed: The rate at which the air is moving.

3 Rivers and valleys

Within the context of rivers and valleys, you should know and understand:

- The three stages of a river: upper course, middle course and lower course.
- The formation of the following landscape features: v-shaped valley, waterfall, meander, ox-bow lake, levée.
- Land uses appropriate to rivers and valleys including farming, forestry, industry, recreation and tourism, water storage and supply, and renewable energy.
- The conflicts which can arise between land uses within this landscape. The solutions adopted to deal with the identified land use conflicts.

You also need to develop the following skills:

- Locate on a map named examples of different rivers and their features in the UK.
- Describe a river and its valley on an OS map.
- Identify landscape features on an OS map, including v-shaped valley, waterfall, meander, ox bow lake, levee.

The location of rivers in the UK

Rivers are part of the **water cycle** and act as natural gutters by collecting all the rain that falls in their surrounding landscape: this is called the **catchment area**. The start of a river is called its **source** and most rivers begin in upland or mountainous areas. As the river flows downhill, it transports water back to the sea, where the river ends: this is called the **mouth** of the river. Look at Figure 3.1, which shows the location of some important rivers in the UK. Notice how they all begin in upland areas and end at the sea.

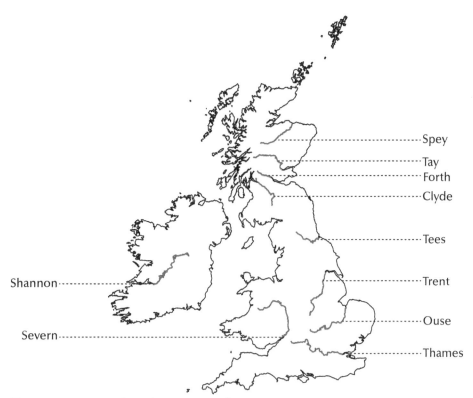

Figure 3.1: *Location of significant rivers in the UK*

River processes

Rivers shape the landscape in three main ways: **erosion, transportation** and **deposition**. **Erosion** is when the river wears away land and the stones carried in it. **Transportation** is the movement of rocks and **silt** by the river. **Deposition** is when the river dumps rocks and silt wherever it slows down because it no longer has the energy to carry its **load**.

Erosion

A river erodes in four ways:

1. **Hydraulic action** – the breaking away of the river bed and banks by the sheer force of the water getting into small cracks and forcing pieces of rock to break off.

2. **Corrasion** – the wearing away of the river bed and banks by the river's load hitting against them and causing the landscape to break up.

3. **Solution/corrosion** – when the water in the river dissolves minerals from the rocks and washes them away.

4. **Attrition** – the wearing down of the load as the rocks hit the river bed and each other, breaking into smaller and more rounded pieces.

Transportation

Look at Figure 3.2, which shows how a river carries its load.

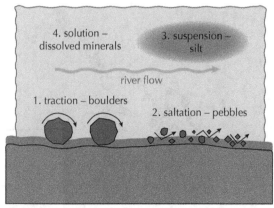

Figure 3.2: *How rivers carry rocks*

The river transports materials in four ways:

1. **Traction** – when large stones e.g. boulders are rolled or dragged along the river bed by the force of the water.
2. **Saltation** – when smaller stones such as pebbles bounce off each other and are carried by the water.
3. **Suspension** – when small particles, e.g. silt, are lifted in the water and carried long distances.
4. **Solution** – when the river dissolves minerals from the rocks that are carried in the water

Stages of a river

A river has **three stages** called the **upper course**, **middle course** and **lower course**. It has specific features at each stage. Figure 3.3 shows the profile of a river and its valley.

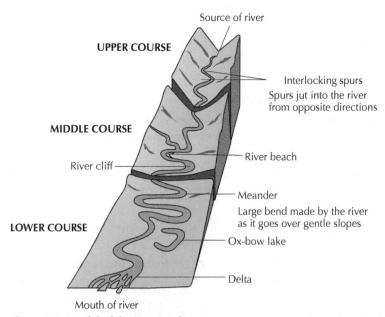

Figure 3.3: *Model of the course of a river*

Characteristics at each stage

Look at Figure 3.4. It lists the characteristics at each stage of the river.

Characteristics	Upper Course	Middle Course	Lower Course
Slope	steep	quite steep	gentle
Width	narrow	quite wide	wide
Depth	shallow	quite deep	deep
Straightness	winding	meandering	large meanders
Amount of load	little	some	lots
Type of load	large/angular	medium and small/ rounded	very small and rounded
Main work of the river	erosion and transportation	transportation	transportation and deposition
Valley width	narrow	quite wide	wide
Main features	v-shaped valley, waterfall, gorge	meander, river cliff, river beach/slip-off slope	flood plain, ox-bow lake, levées

Figure 3.4: *The characteristics at each stage of a river*

The upper course of a river

Landscape	Main Process(es)	Main Features
steep land	vertical erosion (downwards)	• v-shaped valley with interlocking spurs • waterfall • gorge

Figure 3.5: *Characteristics of the upper course of a river*

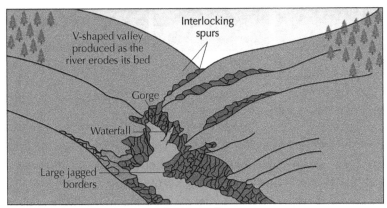

Figure 3.6: *The upper course of a river valley*

The formation of landscape features in the upper course of a river valley

V-shaped valley

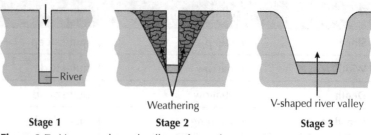

Stage 1   **Stage 2** — Weathering   **Stage 3** — V-shaped river valley

Figure 3.7: *How a v-shaped valley is formed*

- A river flows downhill quickly and erodes the landscape vertically.
- The river cuts a deep gash into the landscape using **hydraulic action, corrasion and corrosion.**
- As the river erodes downwards the sides of the valley are exposed to **freeze-thaw weathering.** This process loosens rocks and they fall into the river. This helps to produce steep valley sides.
- The rocks which have fallen into the river assist the process of **corrasion** which leads to further erosion.
- The river transports the rocks downstream. The process of **attrition** helps to break rocks down and they become smaller and rounder.
- The river channel becomes wider and deeper creating a **v-shaped valley** between **interlocking spurs**.

🔍 HINT

Make sure your diagrams are clearly labelled.

Waterfall

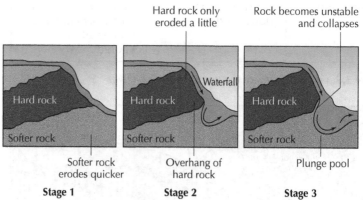

Hard rock only eroded a little   Rock becomes unstable and collapses

Softer rock erodes quicker   Overhang of hard rock   Plunge pool

Stage 1   **Stage 2**   **Stage 3**

Figure 3.8: *How a waterfall is formed*

- The river flows over bands of **hard** and **soft rock**.
- Softer, less resistant rock is quickly eroded by the processes of **hydraulic action**, **corrosion** and **corrasion**.
- The river undercuts the harder rock leaving an **overhang** of **hard rock**.
- The river erodes the softer rock below the **waterfall** by the process of **hydraulic action** and forms a **plunge pool**.

- The overhang of hard rock is unsupported and collapses into the plunge pool below.
- The **waterfall** moves back **upstream** and a **gorge** is cut into the landscape.

Land use in the upper course

- **Hill sheep farming** is the most common type of farming due to poor soils, steep slopes, cold weather, high rainfall and exposed hillsides.
- **Forestry** can be more cost-effective than farming due to the thin soils and steep slopes. Trees can often survive the harsh weather conditions in upland areas.
- **Recreation and tourism: sightseeing waterfalls, gorge walking**, fishing, **canoeing** and **white-water rafting** are all possible activities in the upper course of a river.
- **Reservoirs** can be made, as **dams** can be placed across fast-flowing rivers where v-shaped valleys are narrow because they are easy to block. High rainfall means lots of water can be collected for towns lower down the valley. Where rocks are **impermeable**, **water storage** is easy as the water does not drain away.
- **Hydro Electric Power** (HEP) can be generated in the upper course of a river due to high rainfall and steep slopes providing fast-flowing water to turn turbines and generate electricity.

🔍 HINT

Your diagrams should show the development of a feature over time.

The middle course of a river

Landscape	Main Process(es)	Main Features
moderate – gently sloping **gradient**	**lateral erosion** (sideways) and transportation	• meander • river cliff • river beach/slip-off slope

Figure 3.9: *Characteristics of the middle course of a river*

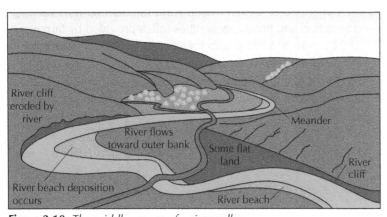

Figure 3.10: *The middle course of a river valley*

The formation of landscape features in the middle course of a river valley

Meander

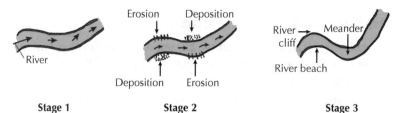

Figure 3.11: *How a meander is formed*

- A river rarely flows in a straight line – it will bend around something in its course, e.g. a tree or hard rock. This results in areas of slower and faster water movement.
- The river contains areas of **deep water** and areas of **shallow water** and this causes the current to swing from side to side.
- The river flows faster on the outside and erodes the outside bends of the river channel. This forms a **river cliff.**
- The river flows more slowly on the inside bend of a river channel and deposits some of its load. This forms a **river beach/slip-off slope.**
- Continuous erosion on the **outer bank** and deposition on the **inner bank** forms a **meander** in the river.
- Over time, meanders become larger and more distinct.

Land use in the middle course

- **Dairy** and **arable farming** are both possible because the land is more gently sloping, soil is more fertile and the weather is warmer and drier than upland areas.

  The flatter land makes it easier for cattle to graze and machines to work in the fields.

- **Recreation and tourism** – roads and railways provide access to all stages of the river and this means **tourists** have access to the upper course for hill walking and sightseeing. They can stay in bed and breakfast facilities (often provided by farmers) and caravan and camp sites located on the gentler slopes in the middle course.

- **Fishing,** e.g. trout and salmon fishing is a good source of income for landowners as they can charge people to fish by selling permts.

 HINT

Remember to include processes of erosion when explaining the formation of river features.

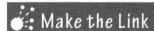 **Make the Link**

In the Rural chapter, you learn about many different types of non-farming activities; B&Bs are just one example of how farmers are diversifying their income.

The lower course of a river

Landscape	Main Process(es)	Main Features
flat land	deposition	flood plain ox-bow lake levées

Figure 3.12: *Characteristics of the lower course of a river*

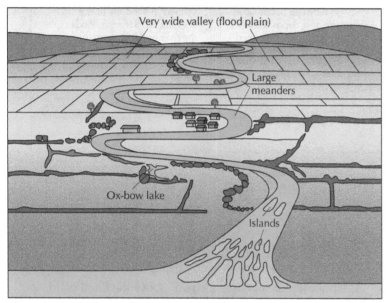

Figure 3.13: *The lower course of a river valley*

The formation of landscape features in the lower course of a river valley

Ox-bow lake

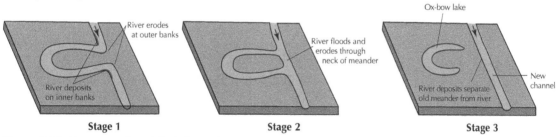

Figure 3.14: *How an ox-bow lake is formed*

- The river is **meandering** across the valley and **erodes laterally** (side of river bank).
- The river flows faster on the outside bends and erodes a **river cliff**.
- The river flows slowly on the inside bends and deposits material forming a **river beach/slip-off slope**.
- Repeated erosion and deposition narrows the neck of the meander.

🔍 HINT

Draw diagrams in pencil so you can easily correct mistakes.

- During a **flood** or intense rainfall, the river will have more energy to erode and it **cuts through the neck of the meander**.
- The river flows on a new, straighter path and the meander is cut-off.
- The river **deposits silt** which seals off the ends of the meander and forms an **ox-bow lake.**

Levée

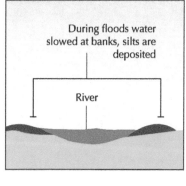

During floods water slowed at banks, silts are deposited — River	In between floods slow moving river deposits silt in riverbed — New levée — New riverbed	With each flood the levées are built up. Between floods the river bed is built up too — Levée — New river level — Levée
Stage 1	**Stage 2**	**Stage 3**

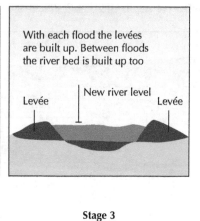

Figure 3.15: *How a levée is formed*

- The river flows on a flat **flood plain** in the lower course.
- The water is flowing slowly and **deposits silt** on the river bed.
- The build-up of silt raises the river above the flood plain.
- When the river floods, a lot of silt is deposited on the river banks.
- **Regular flooding** continues this build-up of silt at the sides of the river.
- The built up **levées (natural embankments)** usually protect the land from further flooding.

🔍 HINT

Use coloured pencils to highlight areas of deposition on your diagrams.

Land use in the lower course

- **Arable farming** takes place due to deep fertile **alluvial** soils on the flood plain. The weather is drier and warmer so crops grow well and more sunshine enables them to ripen before harvest. Machinery can work efficiently as the fields are flat.
- **Dairy farming** can take place as farms are closer to larger **settlements** to sell their produce fresh. The grass is lush for dairy cows to produce high-quality milk.
- **Industry**, e.g. steel works and shipyards, can locate in the lower course due to the large amount of flat land on the flood plain and the proximity to the sea for transporting goods.
- **Recreation and tourism** – sea kayaking and boat trips take place in the lower course. The flat flood plain is used for a variety of buildings including museums and shopping centres.

Make the Link

Think about how the **weather** affects the land use at each stage in the river's course.

⊙ Activity 1: Group activity

1. In groups, choose which stage you want to focus on: upper, middle or lower course.
2. Design a poster highlighting the formations of different river features and land uses in your chosen stage.
3. Create a hand-out for other groups in your class that are working on different stages.
4. Make up a group presentation and present your poster to the class.
5. Give your information leaflet to other groups.
6. Display your poster on the wall along with the other groups' posters.

⊙ Activity 2: Paired activity

1. In pairs, collect small pieces of card and make up revision cards.
2. Draw one feature on the front of each card and write an explanation of how it is formed on the back.
3. Test each other to see if you can explain the formations without using your book or notes in your jotter!

Describing rivers and valleys on Ordnance Survey (OS) maps

When describing a **river** on an OS map, you should include:

- **Direction** – look at the slope of the land.
- **Width** – compare it to its tributaries.
- **Straightness** – is it meandering or fairly straight?
- **Landforms**, e.g. waterfalls, meanders, river cliffs.

When describing a **river valley** on an OS map, you should include:

- **Direction** – look at the contour pattern.
- **Width** – wide or narrow? Look at the distance between the **contour lines** on either side of the river.
- **Slope** – is it a steep v-shaped valley, gently sloping or flat land?
- **Landforms**, e.g. flood plain, levées and ox-bow lakes.

 OS Question

Use the checklist and the OS map to help you describe the middle course of the River Clyde and its valley.

HINT

You must give grid references when answering questions based on an OS map.

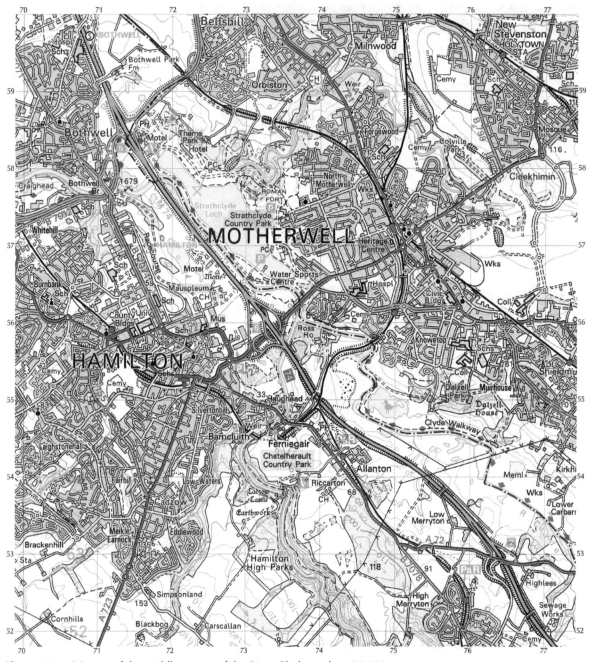

Figure 3.16: *OS map of the middle course of the River Clyde, scale 1:50 000*

Case study of the River Clyde

Introduction

The River Clyde is the third longest river in Scotland. It is about 170 km (106 miles) long and drops 600 m on its journey from its source in the Lowther Hills in the Southern Uplands of Scotland. It travels through many settlements including Lanark, Motherwell, Glasgow and Greenock to its mouth, where it meets the sea at the Firth of Clyde in west central Scotland. The course of the River Clyde is shown in Figure 3.17.

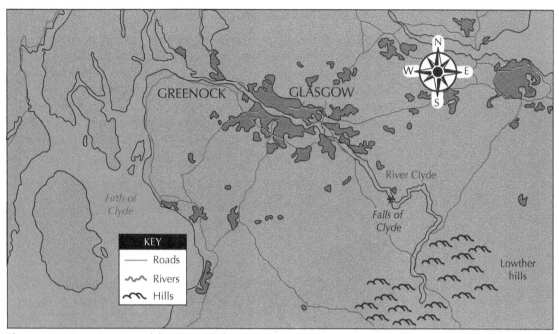

Figure 3.17: *Location map of the River Clyde*

Land use in the upper course

The upper course has famous waterfalls called the Falls of Clyde. People use the paths along the river at this point to sightsee and take photographs of the dramatic river features. **Kayaking** takes place between the waterfalls and fishing is also a popular activity. **Hydro-electric power** is generated using the fast-flowing water at the Falls of Clyde. Daer Reservoir is located close to the source of the River Clyde and provides surrounding settlements with their water supply. The slopes of the valley are steep and mainly used for hill sheep farming and forestry.

Figure 3.18: *The upper course of the River Clyde*

Land use in the middle course

Fishing, e.g. trout and salmon, is common and local angling clubs promote catch and release fishing. The slopes of the river valley are more gently sloping and mixed farming takes place. On the fertile flat land on the valley floor, market gardening is common where fruit, salad vegetables and flowers are grown in large greenhouses. The course of the river has been changed between the towns of Motherwell and Hamilton to create a man-made loch in Strathclyde Park which is used for recreational activities such as sailing.

Figure 3.19: *The middle course of the River Clyde*

Land use in the lower course

Heavy industry traditionally made widespread use of the River Clyde, e.g. docks, ports, water for cooling and dumping waste. The banks of the River Clyde were used for old industries like shipyards and warehouses. With the decline of heavy industries, land use has changed and while shipbuilding remains an important industry, the River Clyde is experiencing massive regeneration, adopting a new identity. Modern recreational facilities such as Intu Braehead Soar, Glasgow Science Centre and the Riverside Museum of Transport are located along the banks of the River Clyde. Commercial facilities include Braehead, the Hydro Music Arena and the BBC. Boat trips are common on the river itself, e.g. on the famous Waverley Paddle Steamer, and Clyde Cruises offer educational trips. Sea kayaking from Glasgow to the Firth of Clyde is also possible.

Make the Link

In the Urban chapter, you learn about land use zones in Glasgow.

Figure 3.20: *The lower course of the River Clyde*

The conflicts which can arise between land uses within this landscape

There are many settlements located along the banks of the River Clyde and people have used it as a source of drinking water, food and transport for centuries. Industries have dammed, straightened, deepened, widened and polluted it. Increased recreation time has enabled people to use the river and the surrounding land for a variety of activities. However, the diversity of land uses of the River Clyde and its valley has resulted in **land use conflicts**.

Conflict between tourists and recreationalists

In the upper course, people want to use the river for different activities. Tourists may want to sightsee, picnic and take photos of the Falls of Clyde (waterfalls) at New Lanark. Local people may want to use the river for recreational activities, e.g. fishing.

Conflict can arise when noisy tourists disturb the peace and quiet for other land users. They may also pollute the river with litter which can negatively affect **wildlife**.

The solutions adopted to deal with the identified land use conflict

- Laws have been passed to protect the River Clyde, e.g. the Salmon and Freshwater Fisheries Act 2003 states that the River Clyde from its source to the Bothwell Bridge is protected water.
- The law states that people using the river and its valley are not allowed to pollute it or damage the wildlife living in it.
- New Lanark is a World Heritage Site which ensures that the river and its surrounding valley are conserved.
- The Scottish Wildlife Trust manages the protection of the **habitats** in the river valley.

HINT

When discussing land use conflicts you should state the two groups involved in the conflict at the beginning of your answer.

HINT

World Heritage Site status ensures that fragile areas are protected for future generations – sustainability is a key concept in Geography.

- Staff at New Lanark promote the Scottish **Outdoor Access Code** to ensure responsible tourism, e.g. promoting the use of litter bins.

Conflict between industry and conservation organisations

In the lower course of the River Clyde water **pollution** is a huge problem due to:

- Chemical **fertilisers** and **pesticides** from farming.
- Pollution from our industrial past, e.g. mining activities.
- Oil from ships using the river for transport.
- **Sewage** from a variety of buildings including shopping centres and offices.
- **Run-off** from landfill sites.

The solutions adopted to deal with the identified land use conflict

- The Scottish Environment Protection Agency (SEPA) has developed a River Basin Management Plan to manage environmental issues within the catchment area.
- SEPA representatives liaise with land managers to monitor and minimise **pollution**.
- SEPA tries to ensure sustainable management and aims to protect the plant community and fish populations and improve water quality in the River Clyde.
- With the relocation of the commercial Port of Glasgow downstream to the deeper waters of the Firth of Clyde, the river has been extensively cleaned up to make it suitable for recreational use.
- The European Union has issued detailed guidance on sewage disposal to minimise pollution levels.
- Glasgow Strategic Drainage Plan focuses on sewerage and drainage management. Some of its objectives are to improve water quality and protect river habitats.

 Activity 3 (National 5)

Individually, choose a river from Figure 3.1 and use the internet to research it. Write a report on your chosen river. You must include:

- Title.
- Location map with labels showing the source and mouth of the river.
- The names of three settlements that the river flows through (marked on the map).
- Information on the land use in each course – upper, middle and lower.
- Details of two land use conflicts.
- Points explaining the solutions to both land use conflicts.
- Suitable pictures to illustrate each section.

Summary

In this chapter you have learned:

- River processes: erosion, transportation and deposition.
- The three stages of a river (upper course, middle course and lower course), their features and characteristics.
- The formation of v-shaped valleys, waterfalls, meanders, ox-bow lakes and levées.
- Land uses appropriate to the three stages of a river and its valley including farming, forestry, industry, recreation and tourism, water storage and supply, and renewable energy.
- Some land use conflicts within the landscape of rivers and valleys and the solutions adopted to deal with these land use conflicts.

You should have developed your skills and be able to:

- Locate on a map named examples of different rivers and their features in the UK.
- Describe a river and its valley on an OS map.
- Identify landscape features on an OS map, including v-shaped valley, waterfall, meander, ox bow lake, levee.

End of chapter questions

National 4 questions

(a) What do erosion, transportation and **deposition** mean?

(b) Name the **three** stages of a river.

(c) Describe the **four** processes of river erosion.

(d) Name the **four** ways that a river transports rocks.

(e) Describe how a river **deposits** rocks it is carrying.

(f) Name **three** features found in the upper, middle and lower course of a river.

(g) What happens to the **width** and **depth** of a river as it moves downstream?

(h) What is the name for the **start** of a river?

(i) What is the name for the **end** of a river?

(j) How does a **river valley** change as the river travels downstream?

National 5 questions

(a) Describe the formation of a **v-shaped river valley**.

(b) Explain how a **waterfall** is formed.

(c) Outline the **use** of a river and its valley in the **upper course.**

(d) Describe the formation of a **meander**.

(e) Outline the **use** of a river and its valley in the **middle course.**

(f) Describe the formation of an **ox-bow lake.**

(g) Explain how **levées** are formed.

(h) Outline the **use** of a river and its valley in the **lower course.**

(i) **Give reasons** for the different land uses located in the upper and lower courses of a river.

National 5 exam-style questions

You can find sample answers to these exam-style questions on the Leckie website:
https://collins.co.uk/pages/scottish-curriculum-free-resources

1

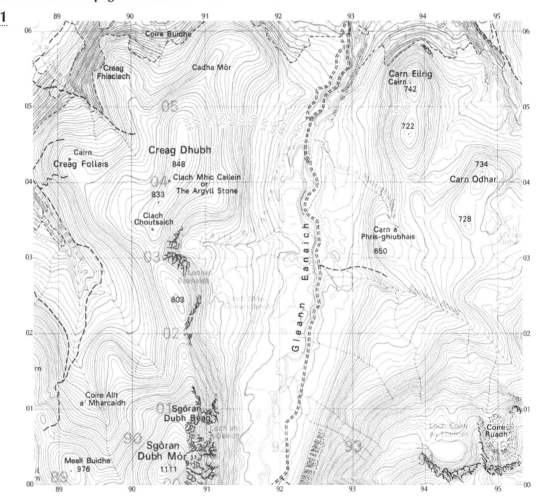

Reference Diagram Q1: *River Einich, scale 1:50 000*

Look at the OS map above.

Describe the physical features of the River Einich **and** its valley from 917000 to 924046.

(5 marks)

2 **Explain** the formation of an ox-bow lake.

(4 marks)

 HINT

Diagrams **must** be labelled and show a change over time.

3

Diagram Q3: *River Forth, scale 1:50 000*

Look at the OS map above.

Describe the physical features of the River Forth **and** its valley from the map extract.

(5 marks)

LEARNING CHECKLIST

Now that you have finished the **Rivers and valleys** chapter, complete a self-evaluation of your knowledge and skills to assess what you have understood. Use traffic lights to help you make up a revision plan to help you improve in the areas you identified as red or amber.

- Describe the terms: erosion, transportation and deposition.

- Explain the processes of river erosion:

  ➢ hydraulic action

  ➢ corrasion

  ➢ attrition

  ➢ solution/corrosion

- List the four ways a river transports rocks.

- Name the three stages of a river.

- State three features at each stage.

- Describe at least three characteristics at each stage, e.g. width, depth, load.

- Label the different features on a diagram of a river's course.

- List the names and meanings of key river features.

- Describe the upper course of a river valley.

- Describe the formation of a v-shaped valley.

- Explain how a waterfall is formed.

- Outline the use of a river and its valley in the upper course.

- Describe the middle course of a river valley.

- Describe the formation of a meander.

- Explain how a river cliff and a river beach are formed.

- Outline the use of a river and its valley in the middle course.

- Describe the lower course of a river valley.

- Describe the formation of an ox-bow lake.

- Explain how levées are formed.

- Outline the use of a river and its valley in the middle course.

- List the conflicts between land uses on the River Clyde.

- Outline solutions adopted to deal with the identified land use conflicts.

- Describe a river and its valley on an OS map.

- Identify v-shaped valleys on an OS map.

- Identify waterfalls on an OS map.

- Identify meanders on an OS map.

- Identify ox-bow lakes on an OS map.

- Identify levees on an OS map.

Glossary

Alluvium: Fine particles of very fertile rock deposited by a river in its lower course.

Arable farming: When farmers grow crops on their land.

Attrition: The wearing down of rocks and pebbles as they hit the river bed and each other, and break into smaller and more rounded pieces.

Canoeing: An activity where people use a small open boat (canoe) to travel on a river.

Catchment area: The area drained by a river.

Contour lines: Orange/brown lines on an OS map which indicate the steepness and shape of the land.

Corrasion: When rocks carried by water wear away the landscape (also called abrasion).

Corrosion: When chemicals in water dissolve minerals in rocks, causing them to break up (also called solution).

Dairy farming: When cows are milked on a farm and the milk is sold to make a profit.

Dam: A large concrete wall built to hold water in a reservoir.

Delta: A landform made of deposited sediment located at the mouth of a river.

Deposition: When a river dumps rocks as the water slows down and no longer has enough energy to transport them.

Embankment: A natural or man-made area of land at the side of a river.

Erosion: The wearing away of land.

Estuary: The area where river water meets sea water when a river enters the sea.

Farming: The growth of plants or rearing of animals for food.

Fertilisers: Chemicals which help crops to grow.

Flood : When a river bursts its banks and overflows

Flood plain: The wide flat valley floor in the lower course of a river.

Forestry: When an area of natural woodland or trees is planted by humans.

Freeze-thaw weathering: When water enters cracks in upland rocks and freezes at night. Continuous freezing and thawing puts pressure on the rocks until small pieces of rock break off.

Gorge: A deep narrow valley with steep rocky sides caused by a waterfall eroding backwards.

Gorge walking: An activity when people walk, swim, climb and jump through the steep sides of a river and its valley.

Gradient: The slope of the land.

Habitats: The areas where plants and animals live.

Hard rock: Rock which is more resistant to erosion.

Heavy industry: Businesses which use bulky raw materials to make heavy goods, e.g. shipbuilding.

Hill sheep farming: When sheep are reared to produce meat and wool.

Hydraulic action: When the power of water forces air into cracks in rocks, compresses it and blows the rock apart as the pressure is released.

Hydro-electric power: Energy which is generated from fast-flowing water.

Impermeable rock: A type of rock which does not allow water to pass through its joints and cracks, e.g. shale.

Industry: The type of work that people do or business activity.

Inner bank: The inside bend of a meander.

Interlocking spurs: The criss-cross appearance of a v-shaped river valley.

Kayaking: A leisure activity where people use a small enclosed boat (kayak) to navigate a river.

Land use conflict: A disagreement over the way an area of land is used.

Lateral erosion: The wearing away of the landscape when a river erodes sideways, creating a river cliff.

Levées: Natural embankments made up of silt which are built up at the sides of a river.

Load: The particles of rock carried by a river.

Lower course: The final section of a river located on flat land.

Meander: A bend in a river.

Meandering: When a river winds its way downstream.

Middle course: The middle section of a river which has sloping land.

Mouth: The end of a river where it meets the sea.

Outdoor Access Code: A policy in the UK designed to educate people about their rights and responsibilities while they are enjoying the great outdoors.

Outer bank: The outside bend of a meander.

Overhang: The piece of hard rock that sticks out over the softer rock in the formation of a waterfall.

Ox-bow lake: A u-shaped body of water which is an old meander cut-off from a river.

Pesticides: Chemicals which kill pests that attack and eat crops.

Physical feature: A natural landform, e.g. meander.

Plunge pool: A hollow at the bottom of a waterfall eroded by the fast-flowing water.

Pollution: When the air, water or land is damaged by harmful chemicals.

Recreation: An activity undertaken during leisure time.

Renewable energy: Energy which is generated from sources that are continually replenished, e.g. wind, wave and solar power.

Reservoir: A man-made body of water used to store drinking water.

River: A natural watercourse which often begins in an upland area and ends when it meets the sea.

River beach: The area where larger pebbles are dropped on the inside bend of a meander (also called a slip-off slope).

River bed: The land at the bottom of a river.

River cliff: The area where fast-flowing water undercuts the valley side and causes it to collapse, leaving a cliff.

Run-off: Water that flows off the land and into a river.

Saltation: When small stones bounce off each other and are carried forward by the water above the river bed for short distances.

Scottish Environment Protection Agency: An organisation responsible for the regulation, protection and improvement of Scotland's environment.

Settlement: A place where people live.

Sewage: Waste water and faeces from toilets.

Sightseeing: A leisure activity when people visit places of interest.

Silt: Fertile sand-sized particles found in the lower course of a river.

Slip-off slope: The area where larger pebbles are dropped on the inside bend of a meander (also called a river beach).

Soft rock: Rock which is less resistant to erosion.

Solution: When particles are dissolved in the river and carried in the water.

Source: The place where a river starts in its upper course.

Suspension: When particles are small enough the river can lift them and carry them long distances.

Tourism: Travel for recreation or business purposes.

Tourists: People who travel for recreation or business purposes.

Traction: When quite large stones are rolled or dragged along the river bed by the force of the water.

Transportation: When a river carries rocks.

Tributary: A smaller river or stream which joins onto the main river.

Upper course: The highest section of a river located in the mountains.

Upstream: The opposite direction to the way a river/stream flows.

Vertical erosion: The wearing away of the landscape when a river cuts downwards.

V-shaped valley: The shape of the landscape which is narrow and steep in the upper course of a river.

Water cycle: The continuous movement of water on, above and below the earth.

Water storage: When water is kept in a place for future use.

Waterfall: A steep drop in a river.

White-water rafting: An activity where people use an inflatable raft and paddles to navigate a river.

Wildlife: Non-domesticated/wild plants and animals.

4 Upland limestone

Within the context of upland limestone, you should know and understand:

- The formation of the following landscape features:
  limestone pavements, potholes/swallow holes, caverns, stalactites and stalagmites, and intermittent drainage.
- Land uses appropriate to upland limestone landscapes including farming, forestry, industry, recreation and tourism, and renewable energy.
- The conflicts which can arise between land uses within this landscape.
- The solutions adopted to deal with the identified land use conflicts.

You also need to develop the following skills:

- Locate on a map named examples of different upland limestone landscapes in the UK.
- Identify landscape features on an OS map including: limestone pavements, pot holes/swallow holes, caverns, stalactites, stalagmites, and intermittent drainage.

The location of upland limestone landscapes in the UK

There are many different types of limestone in the UK but **carboniferous limestone** forms the highest upland areas of them all. Limestone is a hard, **sedimentary rock** formed approximately 350 million years ago. It was formed over a very long period of time when organic matter such as marine organisms built up at the bottom of the sea. Over time, layers upon layers of material built up. This formed the horizontal blocks of rock known as **bedding planes**. As the rock dried out and pressure was released cracks or **joints** were formed. Figure 4.1 shows the location of **upland limestone** landscapes in the UK.

Figure 4.1: *Location of upland limestone landscapes in the UK*

Processes which dissolve limestone

Limestone is a **permeable rock** which means that water can pass through its joints and cracks. It is made of **calcium carbonate** and it is **eroded** by **chemical weathering** – firstly by **carbonation** (when rainwater mixes with carbon dioxide in the air making a weak **carbonic acid**); and secondly by **solution** (when running water carries the limestone away). This causes the limestone to **dissolve** at a very slow rate of about 1 cm every 250 years. The rate at which the limestone dissolves depends on the amount of carbon dioxide in the water. It is the action of water that is responsible for many of the distinctive features of upland carboniferous limestone scenery.

Upland limestone – surface features

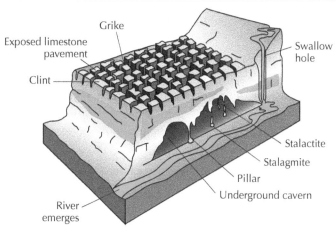

Figure 4.2: *Upland limestone*

Limestone pavements

Figure 4.3a: *A limestone pavement at Malham Cove*

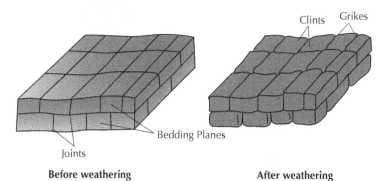

Figure 4.3b

- During **glaciation** (10 000 years ago), ice scraped away the topsoil.

- This process is known as **abrasion** and it exposed the bare rock underneath.
- Today the dry, well-jointed bare rock surface allows water to seep down into it.
- Acidic rainwater (a weak carbonic acid) reacts with the limestone as it passes through the permeable rock.
- It dissolves the rock, enlarging the cracks and making them wider and wider.
- Continued chemical weathering widens and deepens cracks to form gaps called **grikes**. (Figure 4.3b)
- Rectangular blocks of limestone called **clints** are separated by the grikes.
- The resulting pattern of block-like rock is called a **limestone pavement**, e.g. Malham Moor in the Yorkshire Dales. (Figure 4.3a)

HINT

Remember the **g**aps are the **g**rikes.

Swallow holes/potholes

Figure 4.4a: *Gaping Gill Swallow Hole in the Yorkshire Dales*

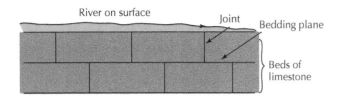

Before weathering

Side view

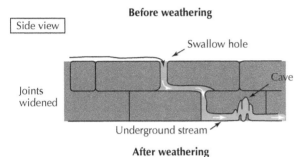

After weathering

Figure 4.4b

- When a stream flows onto limestone it enlarges the joints in the rock. (Figure 4.4b)

- Water seeps down through the rock layers where the joints have been extensively enlarged by chemical weathering.

- Eventually, the stream will disappear down a hole and flow along underground channels.

- The resulting hole in the surface of the landscape is called a **swallow hole**, e.g. Gaping Gill in the Yorkshire Dales. (Figure 4.4a)

- Swallow holes are sometimes referred to as **potholes**.

HINT

Limestone is **p**ermeable which means it allows water to **p**ass through the joints and cracks.

Intermittent drainage

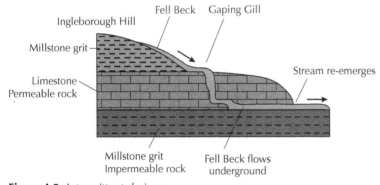

Figure 4.5: *Intermittent drainage*

- If limestone is capped by **impermeable** rock, surface streams can run onto limestone.

- As limestone is permeable, the water disappears from the surface, either by seeping through joints in the limestone or as streams disappearing down swallow holes.

- The underground stream will flow down through the limestone until it reaches an impermeable rock, e.g. millstone grit.

HINT

Millstone Grit is an impermeable rock, so water cannot pass through it.

HINT

Intermittent drainage means that streams appear and disappear irregularly – this can be easily seen on an OS map of the Yorkshire Dales.

- It will emerge at the surface where the water can no longer pass through it to disappear underground.
- An example of a disappearing/reappearing stream is Fell Beck in the Yorkshire Dales.

Scars

Figure 4.6: *Twistleton Scar in the Yorkshire Dales*

Scars are bare limestone cliffs. They were created during the last ice age, when huge ice sheets scraped away the soil-covered sides of valleys.

- Today, the exposed surface on the well-jointed limestone is affected by a process called **freeze-thaw weathering** or **frost shattering.**

- Freeze-thaw weathering occurs when water enters the cracks in the limestone.
- When the temperature drops at night, the water freezes and turns to ice.
- As the ice expands, so do the cracks.
- Repeated freezing and thawing eventually breaks off pieces of rock. At the bottom of the scar cliff a **scree** slope is formed.
- An example of a scar is Twistleton Scar in the Yorkshire Dales. (Figure 4.6)

> **⚡ Make the Link**
>
> Freeze-thaw weathering also occurs in glaciated uplands.

Upland limestone – underground features

Caverns

Figure 4.7a: *Battlefield Cavern in the Yorkshire Dales*

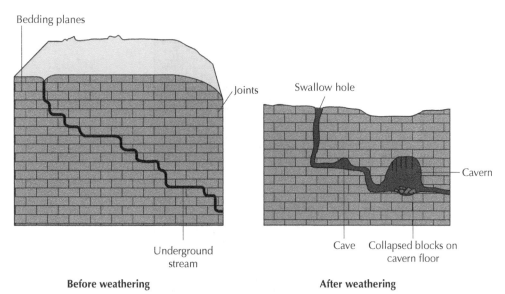

Figure 4.7b

HINT

A cave is much smaller than a cavern.

- Small **caves** develop where joints and bedding planes are close together.
- They form when underground streams are very effective at weathering and eroding the limestone.
- The water dissolves the limestone along the joints and bedding planes more quickly than the rock around it.
- Larger **caverns** form when the roof of a cave collapses.
- They can also be enlarged by the usual processes of river erosion – **hydraulic action, corrosion** and **corrasion**.
- An example of a cavern is Battlefield Cavern in the Yorkshire Dales. (Figure 4.7a)

Inside the cavern there are a number of distinctive features which have formed over thousands of years. They form as a result of water permeating the rock, dissolving the limestone and re-depositing it through a process known as **gas diffusion**. A stalactite is a long, thin, icicle-shaped piece of limestone attached to the roof of a cavern. A stalagmite is a short, dumpy piece of limestone attached to the floor of a cavern. A **rock pillar** is sometimes formed when stalactites and stalagmites join together.

Stalactites and stalagmites

Figure 4.8a: *Stalactites and stalagmites*

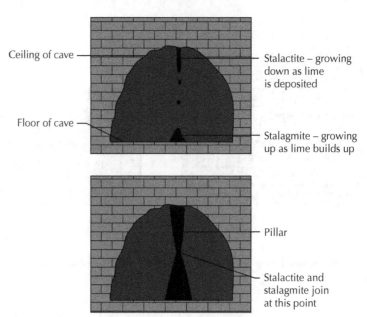

Ceiling of cave — Stalactite – growing down as lime is deposited

Floor of cave — Stalagmite – growing up as lime builds up

Pillar

Stalactite and stalagmite join at this point

Figure 4.8b

- When water flows underground it is loaded with dissolved limestone (calcium carbonate).
- Water drips from the roof of a cavern. When the water **evaporates** tiny amounts of solid calcite are **deposited** on the cavern roof.
- These deposits build up over a very long period of time and form features called **stalactites** which hang down from the ceiling of a cavern.
- An example of a stalactite is the Witch's Fingers inside White Scar Cave in the Yorkshire Dales.
- Some drips of water drop on the floor and some of it splashes and evaporates.
- The splash spreads the deposit of calcite on the cavern floor.
- As more and more calcite builds up, short and dumpy features grow upwards from the ground. These are called **stalagmites**.
- Occasionally stalagmites and stalactites meet to form a **rock pillar**.
- An example of a stalagmite is the Judge's Head inside White Scar Cave in the Yorkshire Dales.

> $\mathcal{P}$ HINT
>
> Stala**c**tites hang from the **c**eiling of a cavern and stala**g**mites grow upwards from the **g**round.

Activity 1: Group activity

1. In groups of four, complete an **expert jigsaw** to discuss the formation of different limestone features.
2. Individually, select a limestone feature by picking lolly sticks/cards.
3. Prepare your mini presentation in your jotter.
4. Present your feature to your group using show me boards.
5. Make a group poster to summarise the surface and underground features of upland limestone landscapes – without using any books or notes. Remember to include **processes** in your answers, e.g. carbonation, and try to give **names of the features** found in the Yorkshire Dales, e.g. **White Scar Cave**.

Identifying limestone features on Ordnance Survey (OS) maps

When identifying limestone features on an OS map, you should include:

- Limestone pavements
- Swallow holes/potholes
- Disappearing/reappearing streams
- Caves/caverns
- Scars

> $\mathcal{P}$ HINT
>
> The names give the location of the features, e.g. Raven Scar.

> $\mathcal{P}$ HINT
>
> You must give grid references and/or named examples in an OS map question.

▲🚩 OS Question

Use the checklist to give different examples of limestone features on the OS map of Malham.

On an OS map of a limestone landscape there are very few rivers and streams because they are not located on the surface of the landscape. Where there is a mixture of limestone and impermeable rocks, streams will disappear where there is limestone and reappear where there are impermeable rocks (intermittent drainage patterns).

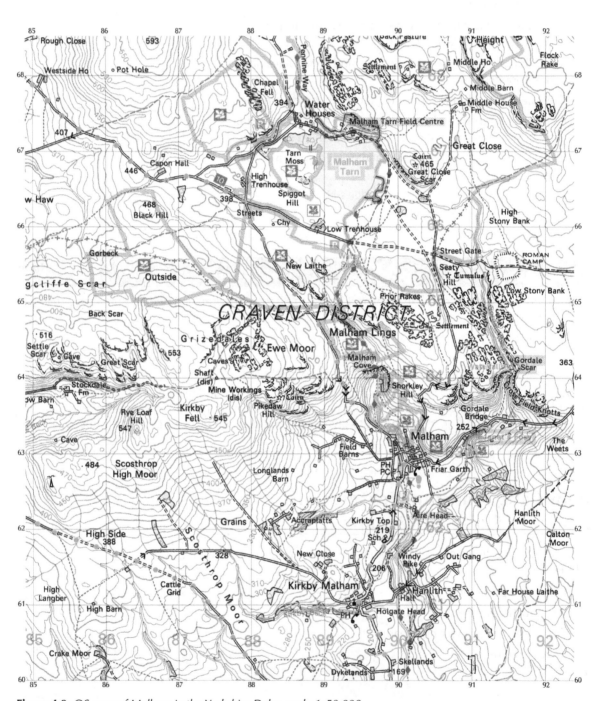

Figure 4.9: *OS map of Malham in the Yorkshire Dales, scale 1:50 000*

Case study of the Yorkshire Dales National Park

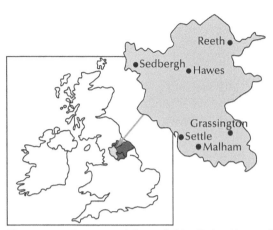

Figure 4.10: *Location of the Yorkshire Dales National Park*

Land uses appropriate to upland limestone landscapes

Farming

Hill sheep farming is the most common type of farming in upland areas. Sheep are hardy and can survive the harsh weather conditions and poor-quality grazing. Some **dairy farms** are located on the flat land in the valleys where the soil is more fertile and the weather is warmer and drier. A lack of surface water, thin soils and bare rock means that crops cannot be grown. Many farmers have **diversified** to non-farming activities to create more income, e.g. bed and breakfast at Dale House Farm in Ingleton.

Figure 4.11: *A hill sheep farm in the Yorkshire Dales*

Make the Link

Hill sheep farming is a common land use in glaciated uplands.

91

Forestry

Upland limestone is largely unsuitable for trees because the limestone is permeable, so there is not enough water for their survival. Forestry plantations are located on millstone grit (an impermeable rock) within the Yorkshire Dales National Park.

Industry

Quarrying in the Yorkshire Dales is an important industry and around 4·5 million tonnes of rock are quarried each year. The main rocks quarried are carboniferous limestone, sandstone and gritstone. Most of the rock is used in the construction industry. One of the largest limestone quarries in the Yorkshire Dales National Park is Swinden Quarry. Cement works also locate in the Dales for the raw material, lime.

Recreation and tourism

Millions of people visit the Yorkshire Dales every year to appreciate the distinctive **scenery**, e.g. Ingleton waterfalls, Malham Cove and Gordale Scar. Visitors enjoy the experience of traditional idyllic rural villages such as Malham. The Dales Countryside Museum enables visitors to find out about the history of the area. People visit limestone caves, e.g. **White Scar Caves** to admire the dripstone feature such as the Devil's Tongue. Ancient semi-natural woodland in Freeholders' Wood has trails for walkers and a variety of **wildlife** to see. Hill walking is a popular activity and there are a number of footpaths including the Pennine Way and Ingleborough Hill. Many other activities such as **caving**, **pot holing**, **rock climbing**, **cycling** and **horse riding** are also popular in the Yorkshire Dales.

Figure 4.12: *Caving in the Yorkshire Dales National Park*

As a result, the Yorkshire Dales National Park is increasingly under pressure to build wind farms. There are currently no large-scale projects as planning applications are always met with widespread opposition from locals. However, it is likely that wind farms will be built in the Yorkshre Dales National Park as the Government have renewable energy targets to meet.

The conflicts which can arise between land uses within this landscape

Many conflicts arise in the Yorkshire Dales National Park as there are so many different land owners and land users. **Two** common conflicts are between: 1. tourists and industry; and 2. locals and tourists. There are various management strategies designed to minimise these conflicts.

Conflict between tourists and industry (quarrying)

Figure 4.13: *A quarry in the Yorkshire Dales*

Tourists visit the Yorkshire Dales National Park to see the beautiful and unusual scenery. Unfortunately, quarries spoil the natural beauty of the landscape, endanger wildlife and put visitors off returning to the area. This threatens local tourism-related jobs, e.g. in shops and restaurants. The large lorries needed to remove the quarried stone cause air **pollution** which spoils the atmosphere for tourists. Lorries also cause **traffic congestion** on narrow country roads which slows traffic and frustrates drivers. The **blasting** of rock creates noise pollution, disturbing the peace and quiet for visitors. Some **wildlife habitats** may also be disturbed by the removal of rock. Quarries can leave a **scarred landscape** when they become disused.

The solutions adopted to deal with the identified land use conflicts

National Park Authorities (NPA) can refuse **planning consents** for new quarries but they cannot close down existing ones. The NPA can insist on companies screening quarries behind trees to reduce visual pollution. Rail transport is encouraged to reduce the number of lorries on roads, e.g. at Swinden Quarry. Lorry loads are also covered to reduce dust in the atmosphere. The movement of trucks and blasting of rocks can be restricted to certain times of the day. Some **nature reserves** are located away from quarries to protect wildlife. Abandoned quarries can make excellent nature reserves and the NPA can insist that quarries are filled in and **landscaped** or turned into a lake to ensure long-term visual pollution is limited.

Conflict between tourists and locals

Figure 4.14: *Part of the Ingelton Waterfalls Trail*

The large volume of visitors means that limestone pavements suffer from wear and tear. Local roads suffer from traffic congestion, particularly near main attractions like Malham Cove. Too many cars on the roads limit local people's movement especially during **peak season**. There is increased air pollution from tourists' car exhausts. Noise pollution from too many people and cars disturbs the peace and quiet in local villages. The demand for car parking often exceeds the number of spaces available which results in cars parking on **grass verges**. Footpaths have been **eroded**, particularly in the areas around Malham Cove and Gordale Scar.

Tourists' litter also spoils the natural beauty of the area. House prices in the villages increase, partly because of incomers wanting to live there and partly because there is a demand for second homes. This means that young people are forced to move away because of the increase in house prices. **Tourist facilities**, signs and man-made walkways make traditional villages look artificial.

The solutions adopted to deal with the identified land use conflict

National Park **legislation** and rules are in place to protect limestone pavements. Park **wardens** help to patrol and monitor the use of footpaths. The use of public transport such as Dales Bus Services is promoted to reduce traffic congestion. Park and Ride services from train stations in larger settlements such as Leeds and Bradford are also encouraged to reduce the number of cars in the Dales. Cycling is allowed on designated cycle paths and Dales Bike Bus which can carry 24 bikes is available. Some farmers open up fields for cars to park during peak season to reduce the number parked illegally. Double yellow lines are also painted on main streets in popular villages. A Dales Visitor Guide which advises visitors on **environmentally positive behaviour** has been published and placed in local tourist information offices. Busy paths have been surfaced and steps have been built up the side of Malham Cove to help prevent further erosion. Litter bins have been removed to encourage visitors to take their rubbish home and dispose of it. The government offers help to first time buyers through Affordable Home Ownership Schemes to enable younger people to stay in their local area. A wider spread of '**honeypot**' areas are being developed to keep tourists away from the most fragile areas. Local authorities can refuse planning permission for new unsightly developments that conflict with one of the aims of the National Park which is to maintain the natural beauty of the area.

🔵 Activity 2

Individually, make a **tourist information leaflet** for visitors to the Yorkshire Dales National Park. You should include:

- Information about the activities and attractions in the area.
- Named examples of places to visit and limestone features to see.
- A paragraph outlining the problems that tourists can cause in the Dales.
- A paragraph on how tourists can minimise their impact on the landscape.
- Colourful pictures and creative writing.

> **GO!** Activity 3 (National 5)
>
> Individually, design and draw a **mind map** summarising the case study of the Yorkshire Dales. You must include:
>
> - Title.
> - Location map of the Yorkshire Dales.
> - Information on the different land uses in the area.
> - Details of two land use conflicts.
> - Points explaining the solutions to both land use conflicts.
> - Suitable pictures to illustrate your mind map.

Summary

In this chapter you have learned:

- Processes which dissolve limestone.

- The formation of limestone pavements, potholes/swallow holes, caverns, stalactites and stalagmites, and intermittent drainage.

- Land uses appropriate to upland limestone landscapes including farming, forestry, industry, recreation and tourism, and renewable energy.

- Some land use conflicts within the landscape of upland limestone and the solutions adopted to deal with these land use conflicts.

You should have developed your skills and be able to:

- Locate on a map named examples of different upland limestone landscapes in the UK.

- Identify landscape features on an OS map including: limestone pavements, pot holes/swallow holes, caverns, stalactites, stalagmites, and intermittent drainage.

End of chapter questions

National 4 questions

(a) Name the features numbered 1–8 on the diagram below. **Choose from**: swallow hole, grike, clint, bedding planes, joint, cavern, impermeable rock and scar.

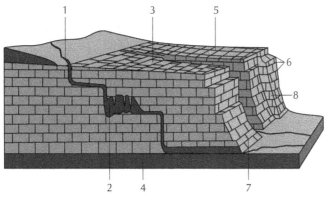

Figure 4.15

(b) What is the difference between bedding planes and joints?

(c) Limestone is a permeable rock. What does that mean?

(d) What is the difference between carbonation and solution?

(e) What are clints and grikes?

(f) What is a limestone scar?

(g) Outline the process of freeze-thaw weathering.

(h) What is the difference between a cave and a cavern?

(i) Describe the process of gas diffusion.

(j) Name **two** features found inside a cavern?

National 5 questions

(a) Describe the formation of a limestone pavement.

(b) Outline the formation of a swallow hole.

(c) Describe how scree is formed.

(d) Explain how caverns are formed.

(e) Describe and explain the formation of stalactites and stalagmites.

(f) State **three common** land uses in upland limestone landscapes.

(g) Describe in detail an example of a **land use conflict** in the Yorkshire Dales.

(h) Outline the **solutions** adopted to deal with the land use conflict described above.

National 5 exam-style questions

You can find sample answers to these exam-style questions on the Leckie website:
https://collins.co.uk/pages/scottish-curriculum-free-resources

A typical exam question will ask you to identify different limestone features on an OS map.

1 Study the OS map extract of Malham in the Yorkshire Dales. (in figure 4.9)

Match the limestone features below with the correct grid references.

LIMESTONE FEATURES: Intermittent drainage/Limestone pavement/Pothole

GRID REFERENCES: 861681/865675/905645/855624

(3 marks)

2 **Explain** the formation of stalactites and stalagmites.

(5 marks)

3 **Describe** the land use conflicts that can occur in an upland area that you have studied.

(6 marks)

4 Outline solutions to the land use conflicts that you identified above. (in question 3)

(6 marks)

LEARNING CHECKLIST

Now that you have finished the **Upland limestone** chapter, complete a self-evaluation of your knowledge and skills to assess what you have understood. Use traffic lights to help you make up a revision plan to help you improve in the areas you identified as red or amber.

- Explain the process of carbonation.

- Explain the process of solution.

- Explain the process of freeze-thaw.

- Label the different features on a diagram of an upland limestone landscape.

- List the names and meanings of key limestone features.

- Describe the formation of a limestone pavement.

- Explain how clints and grikes are formed.

- Describe the formation of a swallow hole/pothole.

- Explain how intermittent drainage works.

- Describe the formation of caverns.

- Explain how stalactites and stalagmites are formed.

- Explain the process of gas diffusion.

- Outline the use of a limestone landscape.

- List the conflicts between land uses in the Yorkshire Dales.

- Outline solutions adopted to deal with the identified land use conflicts.

- Identify limestone pavements on an OS map.

- Identify pot holes/swallow holes on an OS map.

- Identify caverns on an OS map.

- Identify stalactites and stalagmites on an OS map.

- Identify intermittent drainage on an OS map.

Glossary

Abrasion: A type of glacial erosion that happens when rock fragments frozen into the bottom of a glacier scrape and erode the valley floor.

Bedding planes: Layers of sedimentary rock.

Blasting: The process when rock is removed from a quarry.

Calcium carbonate: The main chemical composition of limestone.

Carbonation: When rainwater mixes with carbon dioxide in the air making a weak carbonic acid.

Carbonic acid: When the gas carbon dioxide mixes with water to form an acidic solution.

Carboniferous limestone: A type of sedimentary rock formed between 363 and 325 million years ago.

Cave: A small natural underground space carved out by chemical weathering and running water.

Cavern : A large natural underground space carved out by chemical weathering and running water.

Caving: An activity which involves the exploration of caves.

Chemical weathering: When limestone is eroded by acid rainwater.

Clints: Blocks of limestone in a limestone pavement.

Corrasion: When rocks carried by water wear away the landscape (also called abrasion).

Corrosion: When chemicals in water dissolve minerals in the rocks, causing them to break up (also called solution).

Dairy farming: When cows are milked on a farm and the milk is sold to make a profit.

Deposited: When particles of calcium carbonate are added to an underground limestone feature.

Dissolve: When something is broken up in a liquid and no longer exists.

Diversify: When farmers participate in non-farming activities to generate more income.

Environmentally positive behaviour: Behaviour that causes minimal damage to an area, e.g. when visitors remove their litter.

Eroded: When rock is worn away.

Evaporate: When liquid is vaporised and turns to a gas.

Farming: The growth of plants or rearing of animals for food.

Forestry: When an area of natural woodland or trees is planted by humans.

Freeze-thaw weathering: When water enters cracks in upland rocks and freezes at night. Continuous freezing and thawing puts pressure on the rocks until small pieces of rock break off (also called frost shattering).

Gas diffusion: When calcium carbonate is precipitated out of solution when water evaporates inside a cavern.

Glaciation: A time within an ice age when average temperatures are cold enough for ice sheets to form.

Grass verges: The area of grass located at the side of a road.

Grikes: Gaps in a limestone pavement.

Hill sheep farming: When sheep are reared to produce meat and wool.

Honeypot: An area which has many tourist facilities to attract visitors and keep them away from fragile areas, e.g. Malham.

Hydraulic action: When the power of water forces air into cracks in rocks, compresses it and blows the rock apart as the pressure is released.

Impermeable rock: A type of rock which does not allow water to pass through its joints and cracks, e.g. shale.

Industry: The type of work that people do or business activity.

Intermittent drainage: When water flows over ground on impermeable rock and then disappears underground when it flows onto permeable rock.

Joints: A crack in a rock formed when the rock dries out or pressure is released.

Land use conflict: A disagreement over the way an area of land is used.

Landscaped: An area of land that has been improved by planting grass, trees and other types of vegetation.

Legislation: Laws passed by Parliament.

Limestone pavement: A block-like flat surface of exposed limestone.

National Park: An area that is protected by law to ensure its conservation.

Nature reserve: A protected area where plants and animals are conserved.

Peak season: The busiest time of year for tourism in an area.

Permeable rock: A type of rock which allows water to pass through its joints and cracks, e.g. limestone.

Pillar (rock pillar): A piece of limestone formed by the fusion of a stalactite and stalagmite.

Planning consent: Permission that is needed from the local council before any building work can commence.

Pollution: When the air, water or land is damaged by harmful chemicals.

Pothole: A natural depression on the surface of a limestone landscape eroded by chemical weathering (also called a swallow hole).

Potholing: An activity that involves the exploration of potholes.

Quarried stone: Rock that has been blasted out of the landscape.

Quarry: An area of land that is used to remove rocks for use in industries.

Quarrying: The removal of rocks from the landscape, e.g. limestone and granite.

Recreation: An activity undertaken during leisure time.

Renewable energy: Energy generated from sources that are continually replenished, e.g. wind, wave and solar power.

Scar: An exposed cliff of limestone.

Scarred landscape: When the natural environment has been damaged in some way that makes it look unsightly.

Scenery: The appearance of a place, especially the natural landscape.

Scree: A build-up of broken rock fragments at the bottom of an upland area.

Sedimentary rock: A type of rock formed by the deposition of material at the Earth's surface.

Solution: When running water carries dissolved limestone away.

Stalactites: A long, thin, icicle-shaped piece of limestone hanging from the ceiling of a cavern.

Stalagmites: A short, stumpy piece of limestone growing up from the floor of a cavern.

Sustainable development: When countries use their resources in a way that ensures their use for future generations.

Swallow hole: A natural depression on the surface of a limestone landscape eroded by chemical weathering (also called a pothole).

Tourism: Travel for recreation or business purposes.

Tourist facilities: Services which are provided for visitors, e.g. hotels and souvenir shops.

Traffic congestion: When a large volume of vehicles cause traffic jams.

Upland limestone: A type of sedimentary rock found in mountainous areas.

Upland limestone surface features: Features found on top of a limestone landscape.

Upland limestone underground features: Features found underneath a limestone landscape.

Wardens: People who are employed to monitor an area to ensure that it is being used properly.

Wildlife: Non-domesticated/wild plants and animals.

Wildlife habitats: The places where wild plants and animals live.

Wind power: Renewable energy that is generated by the power of the wind.

5 Glaciated uplands

Within the context of glaciated upland landscapes, you should know and understand:

- The formation of the following landscape features: corrie, pyramidal peak, arête, truncated spur and u-shaped valley.
- Land uses appropriate to glaciated upland landscapes including farming, forestry, industry, recreation and tourism, water storage and supply, and renewable energy.
- The conflicts which can arise between land uses within this landscape.
- The solutions adopted to deal with the identified land use conflicts.

You also need to develop the following skills:

- Locate on a map named examples of different glaciated upland landscapes in the UK.
- Identify and describe landscape features on an OS map including: corrie, pyramidal peak, arête, u-shaped valley and truncated spur.

The location of glaciated uplands in the UK

Glaciation in the UK

> 🔍 HINT
>
> Glaciated uplands are located in mountainous areas.

Although the UK's climate is too warm for glaciers to form today, about 500,000 years ago annual temperatures dropped. Colder temperatures turned rain to snow, which built up on hillsides. Snow turned to ice and over time, large ice sheets formed that were often hundreds of metres thick. Spectacular mountain scenery in the UK, has been carved out by huge glaciers over a series of Ice Ages, the last one ended 10,000 years ago. These glaciers have shaped the landscape in the Highlands of Scotland, Lake District and Northern Wales.

> **Make the Link**
>
> Rivers usually begin in upland areas and end where land meets the sea at the coastline.

Processes of glacial erosion

Glaciated upland landscapes were formed by ice **erosion**. The two processes of glacial erosion are:

1. **Plucking**: ice freezes onto rocks at the side of the landscape and when it moves downhill (under the force of **gravity**), rocks are ripped out. This provides material for abrasion.

2. **Abrasion**: rocks stuck in the base of the ice grind away the landscape underneath it.

> **Make the Link**
>
> Consider the impact of glacial abrasion on upland limestone landscapes.

Upland areas continue to be shaped by **freeze-thaw weathering** today. This occurs when water enters cracks in rocks, freezes and expands. This puts pressure on the crack and forces open a gap. When the ice melts more water can enter the crack and when temperatures drop,

> **Make the Link**
>
> Scree is also formed by the process of freeze-thaw weathering in upland limestone landscapes.

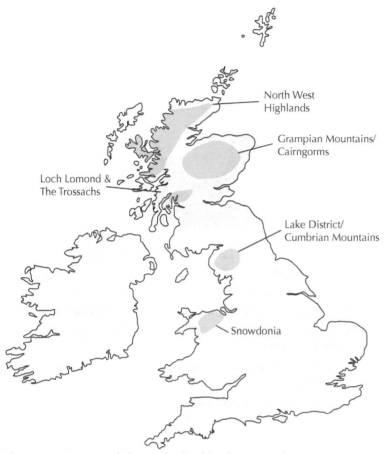

Figure 5.1: *Location of glaciated upland landscapes in the UK*

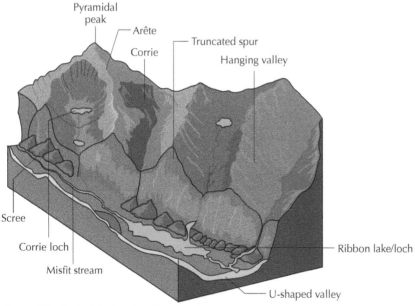

Figure 5.2: *A glaciated upland landscape*

the water freezes again. After many times of freezing and thawing, pieces of rock are broken off the surface. These build up at the bottom of slopes, called **scree**.

Figure 5.3: *Loch Etchachan in the Cairngorms*

Features of erosion in glaciated upland landscapes

Corrie

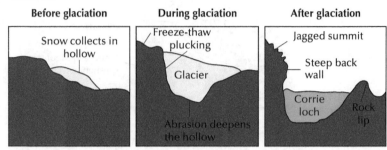

Figure 5.4: *The formation of a corrie*

- Snow gathers in north facing **hollows**. Over time, it builds up and compacts to ice.
- The action of gravity allows the ice to move downhill.
- Ice freezes onto rock on the back wall of the hollow. As it moves, it plucks rock from the landscape. This helps to create a steep back wall.
- Rocks on the back wall are also removed by freeze-thaw weathering.
- Rocks frozen in the base of the ice act like sandpaper and deepen the hollow by abrasion.
- The **rotational movement** of the ice helps to create a deep hollow and a **rock lip** is formed by over-deepening.
- The **glacier** melts and retreats, often leaving a **corrie loch** (or **tarn**) in the base of the **corrie**.
- An example of a corrie in the Cairngorms is Coire Cas.

Arête

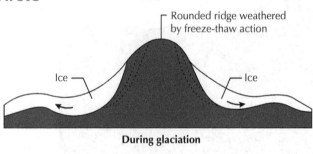

During glaciation

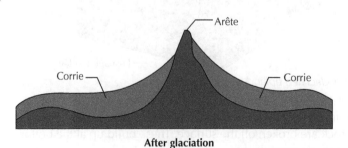

After glaciation

Figure 5.5: *The formation of an arête*

> 🔍 **HINT**
>
> Labelled diagrams are essential when explaining the formation of glacial features.

> 🔍 **HINT**
>
> The formation of physical features should show progression over time, so diagrams must be different at each stage.

- An **arête** is a **ridge** that results when two corries are formed back to back or side by side on a mountain.
- The ice in the corries erodes the hollows using plucking and abrasion.
- As each glacier erodes both corries on either side of the ridge, the edge becomes steeper and the ridge becomes narrower.
- The ridge is jagged as it has been exposed to freeze-thaw weathering.
- An example of an arête in the Cairngorms is Fiacaill Ridge on Corrie Cas.

Pyramidal peak

- A pyramidal peak is formed where three or more corries meet back to back.
- During the ice age snow collected in hollows around a mountain.
- The snow turned to ice, forming glaciers, which moved downhill due to gravity.
- Glaciers erode backwards towards each other removing rocks using the processes of plucking and abrasion. This creates corries on all sides of the mountain.
- Narrow ridges, called arêtes, separate the corries.
- As the glaciers erode backwards into the mountain, the corries get bigger; this produces a steep-sided peak between them.
- Freeze-thaw weathering erodes the top of the mountain and creates a pointed peak.
- An example of a pyramidal peak in the Cairngorms is the Angel's Peak.

U-shaped valley

> 🔍 **HINT**
>
> Try to explain the processes of plucking, abrasion and freeze-thaw weathering.

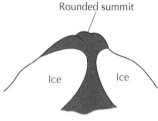

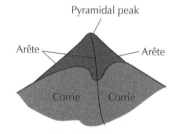

Figure 5.6: *The formation of a pyramidal peak*

> 🔍 **HINT**
>
> Try to describe the process of freeze-thaw weathering in detail when explaining the formation of a pyramidal peak.

> 🔍 **HINT**
>
> Make sure your writing is clear so that it is easy to spot the difference between a **V**-shaped river valley and a **U**-shaped glaciated valley.

Figure 5.7: *Loch Avon in the Cairngorms*

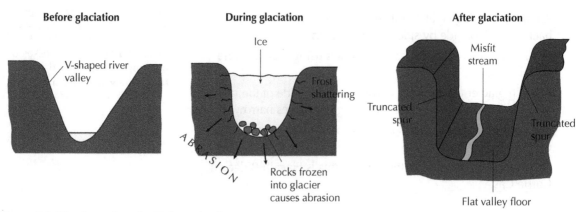

Figure 5.8: *The formation of a U-shaped valley*

Make the Link

Consider how a misfit stream continues to shape the landscape today through the different processes of river erosion.

HINT

Remember to include processes of plucking, abrasion and, if applicable, freeze-thaw weathering in your explanation.

- Snow and ice compress to form a glacier in a v-shaped river valley.
- The glacier uses the process of plucking to steepen the sides of the valley.
- Ice abrasion widens and deepens the valley creating a flat valley floor.
- The valley is also weathered above and below the glacier by freeze-thaw weathering.
- **Interlocking spurs** in the v-shaped valley are cut-off by ice creating **truncated spurs**.
- The 'v'-shaped river valley is changed into a 'u' shape by ice erosion.
- A **misfit stream** or **ribbon loch** (also called a **ribbon lake**) may occupy the valley floor when the ice melts.
- An example of a u-shaped valley in the Cairngorms is Glen Einich.

Activity 1: Paired activity

1. In pairs, make up **revision cards** showing the formations of glacial features.
2. Draw diagrams on the front and write explanations on the back of the cards.
3. Test each other by taking turns to explain the formation of each feature.
4. Choose one feature for your partner and they will choose one for you.
5. Individually take a blank card and, without using the revision cards, books or notes in your jotter, try to draw the diagrams and write how your feature is formed.
6. Once you are both finished, check each other's cards and add in any information that has been missed out.

Identifying glacial features on Ordnance Survey (OS) maps

When identifying glacial features on an OS map, you should include:

- Corries and corrie lochs
- U-shaped valleys and truncated spurs
- Misfit streams and ribbon lochs
- Arêtes
- Pyramidal peaks
- Truncated spur

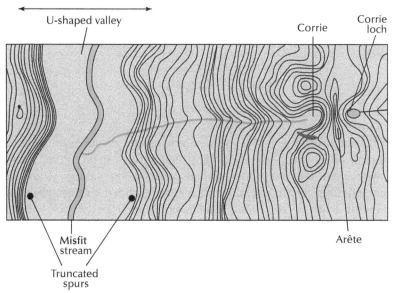

Figure 5.9: *Identifying glacial features on an OS map*

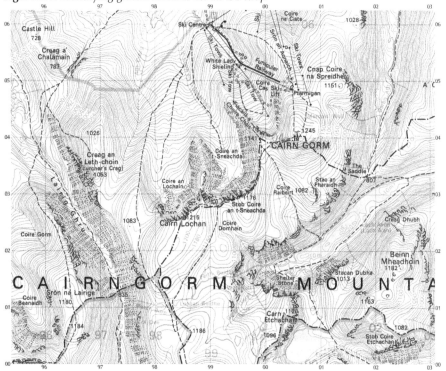

Figure 5.10: *OS map of the Cairngorms*

> 🔍 **HINT**
>
> Names can help to identify features, e.g. **Coire Cas** is a **corrie** in the Cairngorms.

> 🔍 **HINT**
>
> A pyramidal peak is often marked by a spot height (black dot) as it shows the highest point.

> 🔺 **OS Question**
>
> Try to find different examples of glacial features on the OS map extract of the Cairngorms.

Case study of the Cairngorms National Park

Land uses appropriate to glaciated upland landscapes

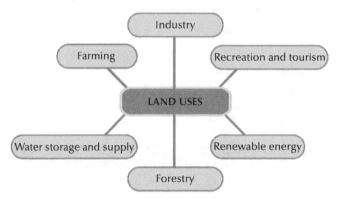

Farming

Hill sheep farming is the most common type of agriculture in an upland area. Sheep are hardy and can survive in the cold, harsh environment on the upper slopes of u-shaped valleys. The low temperatures and lack of sunshine mean the climate is unsuitable for growing crops. High rainfall **leaches nutrients** from the soil, leaving it thin and **infertile**. The slopes are also too steep for using farm machinery. Flatter areas on valley floors are often marshy which makes it unsuitable for arable farming. Pastoral farming is possible on the valley floors. Cattle can graze as the grass is better quality due to fertile soils on the flat land. **Hay** and **silage** are grown as **fodder crops** for animals in winter. **Diversification** also enables hill farmers to earn extra income, e.g. opening camp sites on valley floors for tourists.

> ### Make the Link
>
> Consider the impact of government policies and diversification on the income of hill farmers in Scotland.

> ### Make the Link
>
> Hill sheep farming is also popular in upland limestone landscapes.

Figure 5.11: *Hill sheep farming in the Cairngorms*

Forestry

About 12% of the Cairngorms is forested. **Commercial forestry** takes place on the lower slopes of the u-shaped valleys where conditions are less harsh and soils are better quality. This is possible as trees are hardy and can grow on quite steep land and relatively thin soils. Trees make use of land that is unsuitable for farming or building on and help to prevent soil erosion and flooding. Commercial forestry schemes provide jobs in rural areas. Natural forestry provides a habitat for **wildlife** and encourages tourism by the creation of nature trails and picnic areas. The Queen's Forest is a large area of mixed forest used for many activities including cycling, walking and bird watching. The Forestry Commission organise school visits to various natural attractions including Glenmore forest.

Make the Link

Consider the similar effects of planting trees in the rainforest.

Industry

The main **industry** in the Cairngorms is **quarrying**, e.g. Meadowside Quarry in Kincraig. Slate is extracted and used for roofing buildings and to repair dry stone walls. Granite is quarried and used in different industries such as construction and making roads.

Make the Link

Consider the variety of rock types found in upland and lowland areas.

Recreation and tourism

Tourists are attracted to the **Cairngorms National Park** for its natural scenery which includes ancient forests, vast mountains with glacial features, rivers and lochs. **Ribbon lochs** such as Loch Morlich provide opportunities for water sports such as water skiing and canoeing. Five of Scotland's six highest mountains are located within the Cairngorms National Park, e.g. Ben Macdui. They provide great opportunities for hill walking and rock climbing. Corries such as Corrie Cas enable winter sports such as skiing and snow-boarding. Historical and cultural attractions, e.g. Braemar Castle and ancient ruins, also attract visitors. The Cairngorm Funicular Railway provides year round access to the viewing platform at the top of Cairn Gorm mountain. Some estates offer 4×4 Landrover safaris. Bird watching native species such as the golden eagle and osprey is also popular. Deer stalking, grouse shooting and fishing are common and many of the estates offer permits. The A9 provides access to the National Park and settlements such as Aviemore and Braemar provide tourist services such as hotels, restaurants, cafes, information centres, car parks and equipment hire shops. **Ranger services** offer talks with educational groups and guided tours.

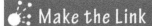

Make the Link

Consider the land use conflict that exists in Figure 5.12.

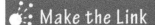

Make the Link

Compare and contrast the activities that take place in both glaciated upland areas and coastal lowlands.

Figure 5.14: *Loch Einich*

Make the Link

Many rural landscapes are suitable for generating renewable energy. Consider the environmental impact of wind turbines and HEP stations.

Figure 5.12: *The Cairngorm Funicular Railway*

Figure 5.13: *Winter sports in the Cairngorms*

Water storage and supply

The high rainfall ensures that lochs in the region can be used to supply drinking water to settlements in the National Park. The hard, **impermeable rocks** provide excellent geological conditions for water storage in **reservoirs**. Steep sided u-shaped valleys provide a natural basin for water storage, e.g. Loch Einich.

Renewable energy

Hydro-electric power (**HEP**) is generated by damming **hanging valleys** to create electricity using the force of water falling. **Wind turbines** can also be located on mountains to take advantage of the windy conditions to generate energy. However, there is continuous conflict between **renewable energy** companies and conservationists. There is a need to generate more 'green' energy while at the same time preserving the natural beauty of the landscape, particularly in protected areas like the Cairngorms National Park.

Figure 5.15: *Wind farms: a scar on the landscape*

The conflicts which can arise between land uses within this landscape

Many conflicts arise in the Cairngorms as there are so many different land owners and land users. Most of this conflict is due to the large number of visitors to the area. **Two** common conflicts are between: 1. locals and tourists; and 2. tourists and farmers. There are various management strategies designed to minimise these conflicts.

HINT

You must know a variety of land uses in a glaciated upland landscape.

Conflict between local residents and tourists

Figure 5.16: *Hill walkers in the Cairngorms*

HINT

When describing land use conflicts you must state the people involved in the conflict.

HINT

The aims of Scotland's National Parks are designed to balance land use and conservation.

Roads become congested due to a high volume of visitors, especially when skiing conditions are good in winter. Narrow roads in small settlements are not built to withstand the sheer volume of cars and parking is also a problem. More traffic increases noise and air pollution. Traffic congestion at peak times prevents local people from going about their daily business. Footpath erosion and litter result in visual pollution in popular walking areas. Fragile wildlife habitats are also destroyed by walkers and wildlife is disturbed in forests and moorland. In summer time, wild camping and fires create litter and increase the risk of accidents. Second home ownership increases in small villages, e.g. Ballater, as more people buy second 'holiday' homes. This causes local first time buyers problems and they may have to move away as houses are unaffordable. Services may close as second home owners are not permanent residents so less money is spent in the area.

The solutions adopted to deal with the identified land use conflicts

Sustainable tourism is promoted to help conserve the Cairngorms National Park for the use of future generations. Ranger services help to promote the understanding, enjoyment and protection of the National Park. The **National Trust** helps to conserve and manage specific areas.

- To protect and enhance the natural and cultural heritage of an area.
- To promote sustainable use of the natural environment.
- To promote public understanding and enjoyment.
- To develop the communities within National Parks, both economically and socially.

Figure 5.17: *The aims of National Parks in Scotland.*

Rail and bus services have been improved and car parks have been built to reduce the number of cars on the roads. Rangers can build and repair stone paths to reduce footpath erosion. Maps, signposts and designated paths help keep walkers off fragile vegetation and direct them along specific routes. There are 46 **Sites of Special Scientific Interest (SSSI)** to help protect native flora and fauna. There are also many information and visitor centres offering education on the various landscapes and diversity of wildlife within the National Park. The government offers help to first time buyers through Affordable Home Ownership Schemes to enable them to buy a property in their local area.

Make the Link

In the Tourism chapter, you will learn about the impact of tourism and ways to manage it in other contexts such as rainforests.

Conflict between tourists and farmers

Tourists can disrupt farming activities: walkers leave gates open and dogs can chase sheep if let off the lead. Stone walls are damaged by people climbing over them. Groups of noisy tourists can disturb animals, especially during breeding season. Farmers may restrict walkers' access at certain times, e.g. lambing season. Farm vehicles can slow up tourist traffic on the roads and parked cars on narrow roads can restrict the movement of large tractors.

Figure 5.18: *A dog loose in a farmer's field*

The solutions adopted to deal with the identified land use conflicts

The Cairngorms National Park Plan was produced to help ensure its sustainable use and conservation. **Park rangers** are employed to prevent problems by encouraging **responsible tourism** and liaise with different land users to minimise problems. Farmers display signs to encourage people to close gates behind them. Voluntary bodies, such as the National Trust, protect areas by buying land and buildings, and maintaining stone walls and footpaths. Visitor centre staff and TV campaigns help to educate the public about the Scottish **Outdoor Access Code** which contains advice about walking dogs within the National Park. Glen Tanar Estate has a website dedicated to promoting responsible dog ownership in the Cairngorms.

Make the Link

Many land use conflicts are similar in glaciated uplands and coastal lowlands.

⊙ Activity 2: Group activity

In groups, investigate and write an illustrated **report** on:

1. The advantages and disadvantages of building wind turbines and HEP stations in the Cairngorms National Park.
2. The problems they cause with other land users.
3. The conflicts they create with the aims of National Parks. Use figure 5.17 to help you.
4. Solutions to these conflicts

You must include a location map and photos to illustrate your findings.

Activity 3 (National 5)

Individually, draw a **mind map** summarising the case study of the Cairngorms. You must include:

- A title and suitable graphics to illustrate your mind map.
- Location map of the Cairngorms National Park.
- Information on the different land uses in the area.
- Details of two different land use conflicts.
- Points explaining the solutions to both land use conflicts.

Summary

In this chapter you have learned:

- Processes of glacial erosion.
- The formation of corries, pyramidal peaks, arêtes and u-shaped valleys.
- Land uses appropriate to glaciated upland landscapes including farming, forestry, industry, recreation and tourism, water storage and supply, and renewable energy.
- Some land use conflicts within glaciated upland landscapes and the solutions adopted to deal with these land use conflicts.

You should have developed your skills and be able to:

- Locate on a map named examples of different glaciated upland landscapes in the UK.
- Describe landscape features on an OS map including: corrie, pyramidal peak, arete u-shaped valley and truncated spur.

End of chapter questions

NATIONAL 4 QUESTIONS

(a) Describe the location of glaciated upland landscapes.

(b) Name the **two** processes of glacial erosion.

(c) What is freeze-thaw weathering?

(d) List **four** features of glacial erosion.

(e) Name the feature often found inside a corrie after the ice has melted.

(f) Match the features numbered on the diagram below to the following words. **Choose from**: u-shaped valley, corrie, hanging valley, arête, pyramidal peak, truncated spur, scree, ribbon loch, misfit stream and corrie loch.

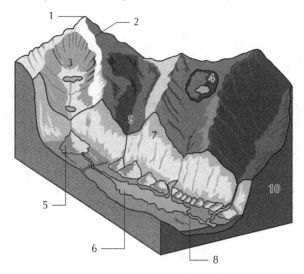

(g) What process makes a pyramidal peak jagged?

(h) Apart from recreation, name **three** different land uses in the Cairngorms.

(i) Outline a land use conflict in the Cairngorms.

(j) Suggest strategies that can reduce the conflict described above.

NATIONAL 5 QUESTIONS

(a) **Describe** the formation of a corrie.

(b) **Explain** how arêtes are formed.

(c) **Describe, in detail,** the process of freeze-thaw weathering in the formation of a pyramidal peak.

(d) Using diagrams, **describe** the formation of a ribbon loch.

(e) **Explain** how ice changes a v-shaped river valley into a u-shaped glaciated valley.

(f) Outline **three** different land users in the Cairngorms.

(g) **Discuss** the conflicts between land uses in the Cairngorms.

(h) **Describe, in detail** the **land use conflict** between tourists and locals in the Cairngorms.

(i) **Outline** the solutions adopted to deal with the land use conflict discussed above.

(j) **Describe** the role of the National Park Authority in managing the Cairngorms.

NATIONAL 5 EXAM-STYLE QUESTIONS

You can find sample answers to these exam-style questions on the Leckie website: https://collins.co.uk/pages/scottish-curriculum-free-resources

1 **Corrie Cas is an example of a glaciated upland feature in the Cairngorms National Park.**

Explain the processes involved in the formation of a corrie, such as Corrie Cas.

You may use diagram(s) in your answer. (4 marks)

2 **Describe, in detail,** land use conflicts that are common in an upland glaciated landscape that you have studied. (5 marks)

3 **Describe** solutions to the land use conflicts identified in your answer above.

(5 marks)

4

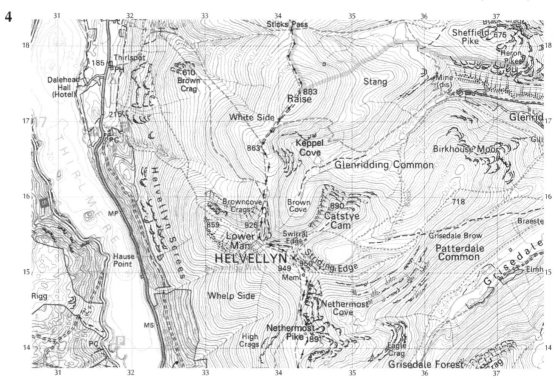

Diagram Q4: *OS map extract of the Lake District, scale 1 : 50 000*

Look at the OS map above.

Identify, using grid references, different glacial features located on the OS map extract.

You must include both rock and water features in your answer. (4 marks)

LEARNING CHECKLIST

Now that you have finished the **Glaciated uplands** chapter, complete a self-evaluation of your knowledge and skills to assess what you have understood. Use traffic lights to help you make up a revision plan to help you improve in the areas you identified as red or amber.

- Locate on a map named examples of different glaciated upland areas in the UK.

- Label the different features on a diagram of a glaciated upland landscape.

- List the names and meanings of key glacial features.

- Explain the processes of:

  ➢ plucking

  ➢ abrasion

  ➢ freeze-thaw weathering

- Describe the formation of a u-shaped valley.

- Explain how corries are formed.

- Describe the formation of arêtes.

- Explain the formation of pyramidal peaks.

- Identify and describe landscape features on an OS map including: corrie, pyramidal peak, arête, u-shaped valley and truncated spur.

- Outline the use of a glaciated upland landscape.

- List the conflicts between land users in the Cairngorms

- Outline solutions adopted to deal with the identified land use conflicts.

Glossary

Abrasion: A type of glacial erosion that occurs when rock fragments frozen into the bottom of a glacier scrape and erode the valley floor.

Arête: A knife-edge ridge between two corries.

Commercial forestry: When trees are grown for sale.

Corrie: A large hollow in the side of a mountain that was carved out by ice.

Corrie lip: A raised area of rock at the front of a corrie.

Corrie loch: A small lake found inside a corrie (also known as a tarn).

Diversification: When farmers undertake non-farming activities to make more money.

Erosion: The wearing away of the landscape.

Fodder crop: Crops such as hay and silage grown to feed animals during winter months.

Freeze-thaw weathering: When water enters cracks in upland rocks and freezes at night. Continuous freezing and thawing puts pressure on the rocks until small pieces of rock break off (also called frost shattering).

Glaciated upland: A mountainous area that has been eroded by ice.

Glaciation: A time within an ice age when average temperatures are cold enough for ice sheets to form.

Glacier: A very large body of ice formed over 30–40 years when snow is compressed by its own weight.

Gravity: A natural force which causes the downward movement of a glacier on a mountain.

Hanging valley: A smaller valley which is located high above the main u-shaped valley.

Hay: Grass that has been cut, dried and baled.

Hill sheep farming: When sheep are reared to produce meat and wool.

Hollow: The dip in a mountain where snow and ice gather.

Hydro-electric power: Energy which is generated from fast-flowing water.

Impermeable rocks: Rocks which do not allow water to pass through them, e.g. granite.

Industry: The type of work that people do or business activity.

Infertile: When soil no longer has the required nutrients for crops to grow.

Interlocking spurs: The criss-cross appearance of a v-shaped river valley.

Land use conflict: A disagreement over the way an area of land is used.

Leach: When rainwater washes all the goodness out of top soil.

Misfit stream: A small river occupying the floor of a u-shaped valley.

Moraine: Rock that has been eroded and carried by a glacier.

National Park: An area that is protected by law to ensure its conservation.

National Trust: A charity responsible for ensuring the protection and preservation of historic places and spaces for public enjoyment.

Nutrients: Chemical elements that are essential for plant nutrition.

Outdoor Access Code: A policy in the UK designed to educate people about their rights and responsibilities while they are enjoying the great outdoors.

Park ranger: A person who works to protect the natural environment.

Plucking: A type of glacial erosion that occurs when ice freezes onto the landscape and when it moves, it rips out rocks.

Pyramidal peak: A steep, jagged point on a mountain top.

Quarrying: The removal of rocks such as slate and granite from the landscape.

Ranger services: People who are employed to monitor and educate the public on the proper use of an area.

Recreation: An activity undertaken during leisure time.

Renewable energy: Energy which is generated from sources that are continually replenished, e.g. wind, wave and solar power.

Reservoir: A man-made body of water used to store drinking water.

Responsible tourism: When visitors are encouraged to cause minimal damage to an area.

Ribbon loch: A large narrow loch occupying the floor of a u-shaped valley (also called a ribbon lake).

Ridge: A long and narrow chain of mountains.

Rotational movement: The circular movement of a glacier inside the hollow of a corrie.

Scree: A build-up of broken rock fragments at the bottom of an upland area.

Silage: Grass that has been cut, baled and wrapped in black cellophane to keep it moist.

Site of Special Scientific Interest (SSSI): A designated area in the UK which is protected.

Summit: The very top of a mountain.

Tarn: A small lake found inside a corrie (also known as a corrie loch).

Tourism: Travel for recreation or business purposes.

Tourists: People who travel for recreation or business purposes.

Truncated spur: A rounded area of land at the edge of a u-shaped valley.

U-shaped valley: A glacial trough that was formed by ice erosion.

Wildlife: Non-domesticated/wild plants and animals.

Wind farm: An area of land containing many wind turbines.

Wind turbine: A large wind mill used to generate energy from moving air.

6 Coastal landscapes

Within the context of coastal landscapes, you should know and understand:

- The formation of the following landscape features: cliffs, caves and arches, stacks, headlands and bays; spits and sand bars.
- Land uses appropriate to coastal landscapes including farming, forestry, industry, recreation and tourism, water storage and supply, and renewable energy.
- The conflicts which can arise between land uses within this landscape.
- The solutions adopted to deal with the identified land use conflicts.

You also need to develop the following skills:

- Locate on a map named examples of different coastal landscapes in the UK.
- Identify and describe coastal landscape features on an OS map including: cliffs, caves, arches, stacks, headlands, bays, spits and sand bars.

The location of coastal landscapes in the UK

There are many different types of coast in the UK. The Dorset (Jurassic) coast is 95 miles long and has a variety of **coastlines** including **cliffs** (Purbeck Cliffs), **headlands** (Peveril Point), **bays** (West Bay) and **beaches** (Bournemouth). Coastal areas were formed in various ways according to the **geology** of the area. Coastlines continue to change depending on **waves** being **constructive** or **destructive**. Constructive waves build up coastlines such as beaches, e.g. Chesil Beach. Destructive waves remove material from the coastline, e.g. Old Harry Rocks. Look at Figure 6.1 which shows the location of coastal landscapes in the UK.

HINT

Coastal areas are located at the edge of a map as a coast is where land meets the sea.

Make the Link

Coastal areas are located in lowland areas whereas glaciated landscapes are found in upland areas.

■ COASTS

Figure 6.1: *Location of coastal landscapes in the UK*

Processes of coastal erosion

Coastal landscapes are formed by **erosion**, **transportation** and **deposition**. The power of the sea shapes the coastal landscape. Waves get their energy from the wind. The size of the wave is determined by: the speed of the wind, the length of time the wind has been blowing and the distance the wind blows over the sea, called the **fetch**. The stronger the waves are, the greater the erosion.

The four processes involved in coastal erosion are:

1. **Hydraulic action.** This is the force of waves crashing against the shore and cliffs. The power of the waves forces air into cracks and compresses it. This compression breaks the rocks apart as the pressure is released.

2. **Abrasion/corrasion.** This is when rocks carried by the waves are thrown against cliffs breaking them up.

3. **Solution/corrosion.** This is when chemicals in seawater **dissolve** minerals in the rocks.

Figure 6.2: *Durdle Door on Dorset coast*

4. **Attrition**. This happens when rocks carried by the waves smash into each other, wearing each other away and gradually becoming smaller, rounder and smoother.

Coastal landscapes – features of coastal erosion

Cliffs and wave-cut platforms

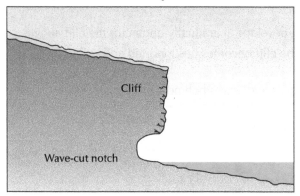

Stage 1: The sea erodes a wave-cut notch

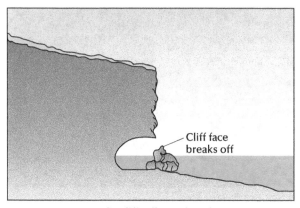

Stage 2: The cliff collapses into the sea

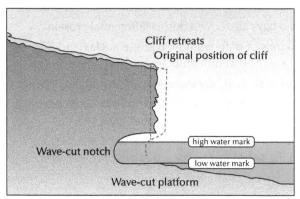

Stage 3: A wave-cut platform marks the remains of a cliff

Figure 6.4: *The formation of a wave-cut platform*

Figure 6.3: *Purbeck Cliffs and Old Harry Rocks*

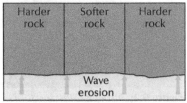

Stage 1: Waves attack coastline.

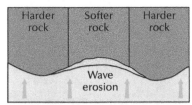

Stage 2: Softer rock is eroded quickly.

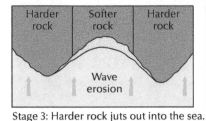

Stage 3: Harder rock juts out into the sea.

Figure 6.6: *Development of headlands and bays*

- Cliffs form where there is harder, more **resistant rock** such as granite or chalk.
- Erosion is greatest where waves break against the base of the cliff and when there is no beach to absorb the wave energy.
- The cliff is **under cut** by the erosive power of the waves by the processes of hydraulic action, abrasion and corrosion. This forms a **wave-cut notch**.
- Continued erosion causes the notch to become larger over time.
- As the notch develops, it gradually undercuts the cliff above.
- Eventually, the cliff becomes unsupported and collapses into the sea.
- The cliff retreats and leaves behind a wave-cut platform.

Headlands and bays

Figure 6.5: *Swanage Bay and Ballard Point*

- The geology of a coastline determines the rate of its erosion.
- Headlands and bays are formed due to **differential erosion**.
- **Alternating bands** of different rock types, e.g. **clay** and **limestone**, are eroded at different rates.
- Clay is a softer rock than limestone so is more quickly eroded.
- The softer rocks erode backwards faster to form sheltered bays that often contain beaches, e.g. Swanage Bay.
- The harder limestone areas are more resistant to erosion and stick out into the sea to form exposed headlands, e.g. Peveril Point.

Caves, arches and stacks

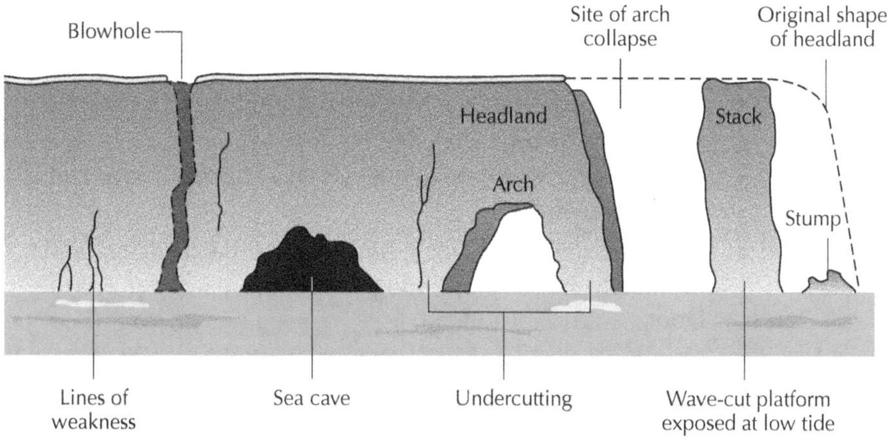

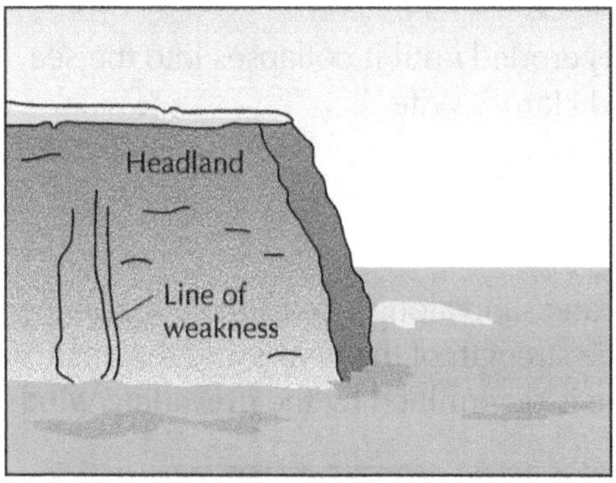

Stage 1: A line of weakness in a headland

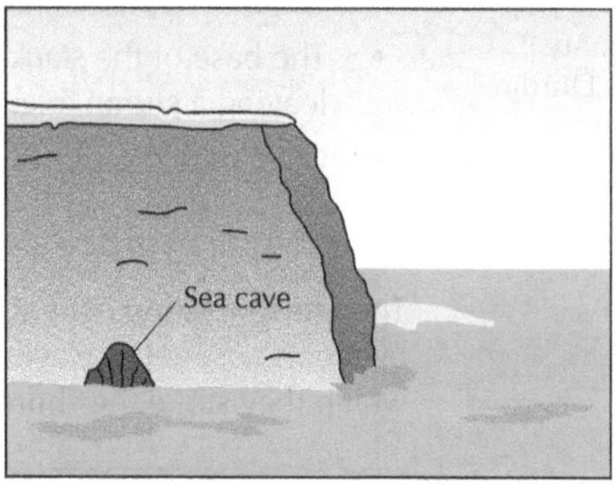

Stage 2: A sea cave is formed

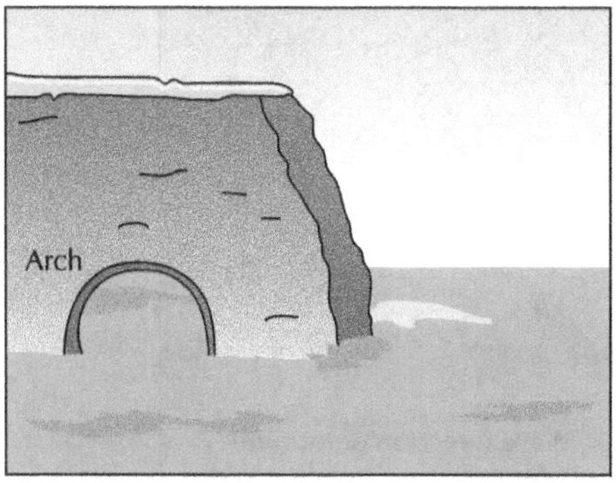

Stage 3: An arch opens up in the headland

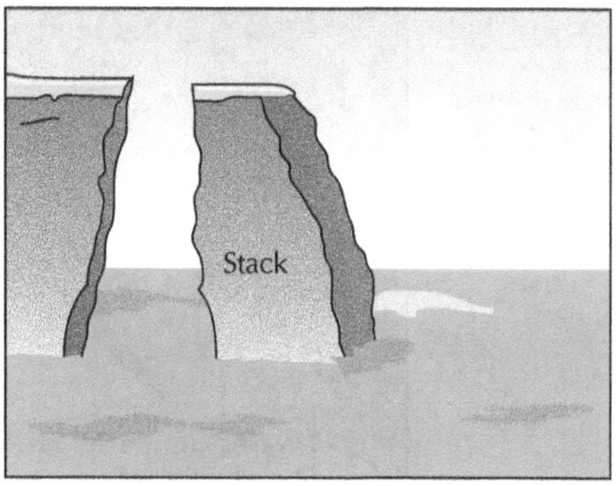

Stage 4: A free-standing stack

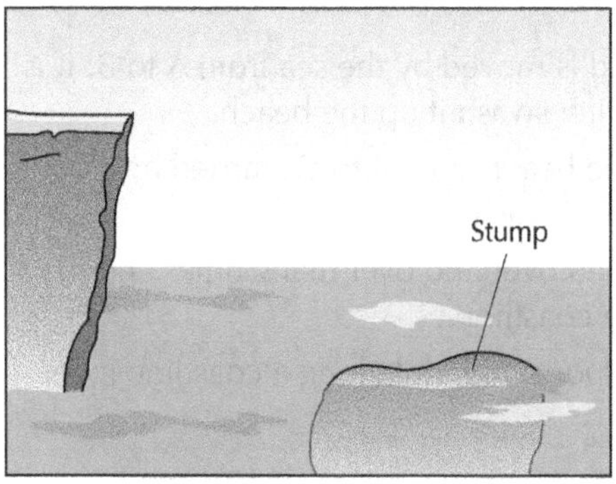

Stage 5: A stump remains

Figure 6.7: *The formation of caves, arches and stacks*

HINT

Think about the processes of erosion taking place.

HINT

The formation of a stack is a sequential process.

- Even resistant rocks contain weak points that are more easily eroded.
- The waves erode along **lines of weakness** by hydraulic action, abrasion and corrosion.
- Over time, lines of weakness are enlarged and develop into small sea caves, e.g. Tilly Whim Cave.
- Occasionally, waves will erode vertically inside a cave and form a **blowhole**.
- The caves are expanded on both sides of the headland until eventually the sea cuts through, forming an arch e.g. Durdle Door.
- Continued erosion enlarges the arch until the rock at the top becomes unsupported.
- Eventually the arch collapses to form a stack, e.g. Old Harry, which is separated from the headland.
- The base of the stack gets eroded until it collapses into the sea leaving a **stump**, e.g. Old Harry's Wife.
- Over time the stump will disappear.

HINT

Try to learn examples of the different features, e.g. an example of an arch is Durdle Door.

Coastal transportation

The **transportation** of pebbles and sand along a coastline is influenced by the movement of waves. The strength of the waves and the angle at which they strike the **shoreline** is determined by the **prevailing wind**.

HINT

The prevailing wind is the most common wind direction.

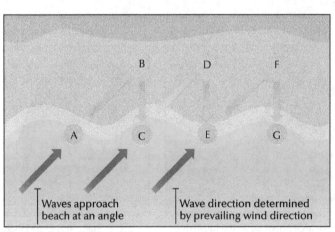

Figure 6.8: *Longshore drift*

- A pebble or grain of sand is moved by the sea from A to B: it is carried by the waves which **swash** it up the beach.
- It is then pulled down the beach from B to C, carried by **gravity** and the wave's **backwash**.
- This movement is repeated over and over many times and the particle travels along the coastline.
- This process which transports material along a coastline is called **longshore drift**.

HINT

Swash sweeps the sand up the beach and **b**ackwash pulls it **b**ack out to sea.

Activity 1: Group activity

1. In groups, investigate:
   (a) Why the coastline is being eroded.
   (b) The methods of protecting the coastline including groynes, rip-rap, sea walls and beach stabilisation.
   (c) The reasons for protecting the coast from erosion.

2. You must include text and photos to illustrate your findings.

3. You should display the information your group has gathered in a PowerPoint presentation.

4. Decide who will talk about each slide and practise your presentation.

5. Present your group PowerPoint presentation to your class.

Figure 6.9: *Chesil Beach*

Coastal landscapes – features of coastal deposition

The formation of sand spits

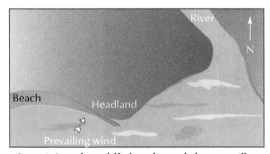
Stage 1: Longshore drift deposits sand along coastline

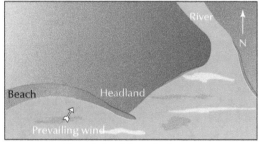
Stage 2: Sand deposited out at sea

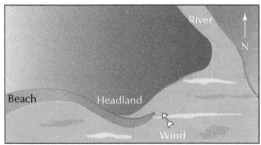
Stage 3: Wind changes direction and alters deposition

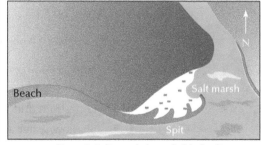
Stage 4: Salt marsh forms behind spit

Figure 6.10: *The formation of sand spits*

- Spits form where the coastline changes direction and longshore drift continues to transport sand.
- The transported sand is deposited in the water that is sheltered by the headland.
- The level of sand will build up until it is eventually above sea level.

- Continued deposition allows the beach to extend into the sea and a spit is formed.
- The spit cannot develop right across the bay as a river's **estuary** prevents deposition.
- Sand spits often have a curved end as a secondary wind and wave direction curves the end of the spit.
- The spit creates an area of calmer water behind it, where a **lagoon**, **salt marsh** or dry land can develop.
- An example of a sand spit on the Dorset coast is Studland Spit.

HINT

The formation of a feature should show a change over time so your diagrams must be different at each stage.

The formation of a sand bar

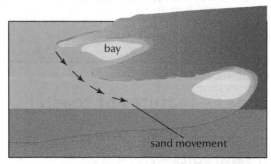

Stage 1: Longshore drift deposits sand along coastline

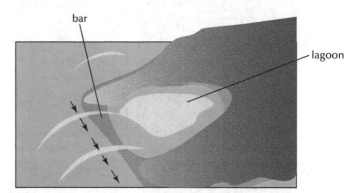

Stage 2: sand bar forms where sand is deposited across a bay

- Where there is no river estuary, the sand and silt can be deposited faster than it is deposited.
- Longshore drift continues to move sand across the bay.
- Shallow water allows deposited sand to build up.
- When the sand extends right across the bay, a sand bar is formed.
- Behind the sand bar a pool of stagnant water called a lagoon will form.
- An example of a sand bar and lagoon is Chesil Beach and the Fleet lagoon in Dorset.

Identifying coastal features on Ordnance Survey (OS) maps

When identifying coastal features on an OS map, you should include:

- Cliffs and wave-cut platforms
- Caves and arches
- Stacks and stumps
- Headlands and bays
- Beaches and spits
- Sand bars

HINT

Sometimes the names give the location of the features, e.g. Swanage Bay.

The table below gives you some clues to look out for when identifying coastal features on an OS map.

Coastal feature	Clue to help identify the features on an OS map
Cliff	The word 'cliff', e.g. West Cliff. The contour lines may appear to run into the sea, indicating the height of the cliffs.
Wave-cut platform	The flat rock symbol on the seaward side of the coastline indicates a wave-cut platform.
Cave	The word 'cave/s', e.g. Tilly Whim Caves.
Arch	The word 'arch', e.g. Natural Arch.
Stacks and stumps	Off the headland there may be small islands; these will be former parts of the headland.
Headland	The word 'point'.
Bay	The word 'cove' or 'bay'.
Beach	Look for the peach colour beside the water. The word 'Sand dunes'.
Spit	Sand extends out into the water and, if you look closely, you should see their **curved hook.**
Sand bar	Sand extends out into the sea and right across a bay. Look for a pool of water behind the sand bar, known as a lagoon.

HINT

You must give grid references and/or named examples in an OS map question.

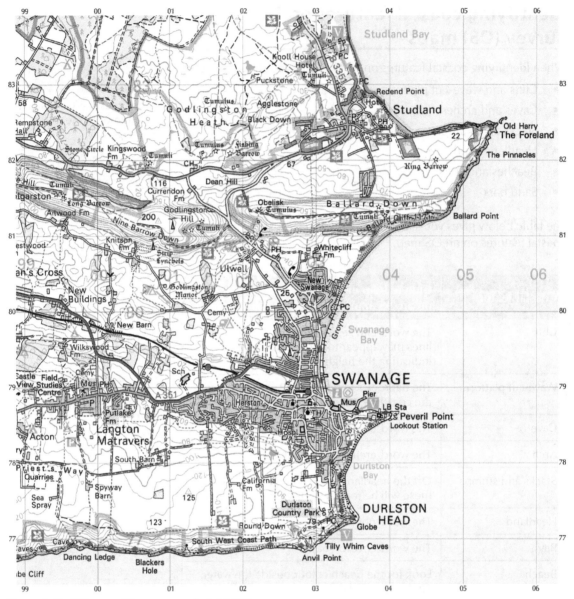

Figure 6.11: *OS map of Swanage, scale 1:50 000*

 OS Question

Try to pick out examples of the coastal features on the OS map of Swanage above.

Case study of the Dorset coast

Land uses appropriate to coastal landscapes

Farming

Arable and **pastoral farming** take place on the flat land behind the coastline, e.g. Greenlands Farm. The warm, sunny climate and fertile clay soils are suitable for growing crops. Animals can also graze on the lush grass. Some farmers have also expanded their income by moving to non-farming activities, e.g. Ulwell Farm Caravan Park. **Oyster farming**, **mussel farming** and fishing take place at Poole Harbour.

Forestry

Although forestry tends to be located on steep slopes and coastal areas are flat, there is a large coniferous forestry plantation in the area behind Studland Bay. Goathorn Plantation is situated on gently sloping land but the land is unsuitable for building on as it has many streams and it borders an area of marshland and mud.

Industry

There are different types of industries located along the Dorset coast: oil extraction takes place at Britain's sixth largest **oil field** located at Poole Harbour. High-quality limestone is quarried on the Isle of Portland. Poole Harbour provides direct access to the English Channel, which is a major shipping route. Here, imports enter the UK and exports are transported to Europe and beyond. Bournemouth Conference Centre is a service industry that provides year-round jobs.

Recreation and tourism

The Dorset coast has many coastal resorts such as Swanage, Lyme Regis and Bournemouth which provide tourist services including hotels, caravan sites, souvenir shops and restaurants. The warm, sunny climate and sandy beaches enable sunbathing and beach sports such as volleyball. Water sports are also popular and include swimming, **water skiing** and **yachting**. There are a number of historic attractions which bring visitors to the Dorset coast including Maiden Castle, Thomas Hardy's cottage and many historical ruins such as the White Horse. Activities also include **fossil hunting** and the geology of the area enables the study of different rock types as the Dorset coast has rocks dating back 185 million years. The area also has many **coastal footpaths**, e.g. the South West Coast Long Distance Footpath where walkers can enjoy the beautiful **coastal scenery**. People come to visit and photograph features such as Lulworth Cove and Durdle Door. The coastline has a variety of **wildlife** due to the diverse landscapes such as sand dunes, lagoons, salt flats and bays. Durlston Head is famous for **bird watching** and Brownsea Island Nature Reserve is the habitat of red squirrels and wildfowl. Chesil Beach is a **sand bar** and the Fleet Lagoon behind it is an important **Site of Special Scientific Interest** due to the diversity of wildlife.

Water storage and supply

There are several **reservoirs** built in the area, including one located behind the small village of Studland. These provide nearby settlements occupied by both local residents and tourists with a fresh, clean water supply.

Renewable energy

There is the potential to generate electricity through wind and wave power. Off-shore wind farms provide a source of **renewable energy** but conflict with the natural beauty of the area. **Nuclear power stations** are often sited on coastal areas as there is plenty of sea water to cool the reactors. Renewable energy sources continue to be a topic of

> ### 🔍 HINT
>
> A sand bar is formed when longshore drift continues out across a bay and cuts off the bay to form a lagoon such as 'The Fleet' lagoon behind Chesil Beach.

> ### 🔍 HINT
>
> You must know a variety of land uses along a coastline – you cannot just focus on tourism.

> ### Make the Link
>
> Consider how land uses differ between coastal landscapes and upland glaciated areas.

debate on the Dorset coast as the area is designated as a World Heritage Site.

The conflicts which can arise between land uses within this landscape

Many conflicts arise on the Dorset coast as there are so many different land owners and land users. Most of this conflict is due to the mass influx of tourists to the region, especially during summer months. **Two** common conflicts are between: 1. locals and tourists; and 2. tourists and other tourists. There are various management strategies designed to minimise these conflicts.

Conflict between local residents and tourists

Figure 6.12: *Weymouth beach in summertime*

Figure 6.13: *Traffic congestion on the A351 coastal road in Dorset*

As there is no motorway access, small coastal roads become easily congested due to the high volume of visitors from large cities such as London and Birmingham. Many tourists also increase congestion in 'honeypot' areas such as Corfe Castle. The higher volume of traffic increases noise and air pollution. The huge numbers of visitors create unsightly footpaths and erosion of fragile landscapes such as Studland sand dunes. Tourists often leave litter, especially on beaches such as Chesil Beach. Fires are common on beaches as people bring food to barbeque. This increases the risk of accidents and can burn the grassy areas along the coastline. Fragile **wildlife habitats** are also destroyed by walkers and wildlife is disturbed on beaches. Second home ownership increases as more people buy second 'holiday' homes. This causes local first time buyers problems getting on the property ladder and they may have to move away as houses are unaffordable.

The solutions adopted to deal with the identified land use conflicts.

The Dorset coastline is protected in various ways by lots of different organisations. **World Heritage Site status** helps to ensure that the coastline and its wildlife are protected. Train lines, buses, cycle routes and boat trips offer alternative forms of transport to reduce **traffic congestion** on coastal roads. A preserved steam railway line from Swanage to Corfe allows sightseers access to this area and decreases the number of drivers on the roads. The **National Trust** and Dorset Wildlife Trust buy and help to manage land, e.g. surface footpaths. Local authorities fine people for littering or dumping. Voluntary **litter picks** are also arranged by local groups. Beach wardens help to manage tourists on the beach and in the sea. **Marram grass** has been re-planted to help **conserve** the vegetation and **sand dunes** at Sites of Special Scientific Interest, e.g. Studland. **Nature reserves** such as Townsend Nature Reserve have been created to protect wildlife. The government offers help to first time buyers through Affordable Home Ownership Schemes to enable them to buy a property in their local area.

Figure 6.14: *Different watersports can cause conflict*

Conflict between tourists and other tourists

Figure 6.15: *Poole Harbour*

There are many different land users in Poole Harbour and approximately 4000 boats use the area during peak summertime. Tourist facilities such as car parks, caravan sites and marinas spoil the natural beauty of the coast, especially for those who come to admire the unusual coastal features. Sunbathers and swimmers may be disturbed by the noise of jet skis and motor boats. People fishing at various coastal spots may be disrupted by activities such as water skiing. Over 20 000 visitors can visit Studland beach on a hot summer's day. This makes the area overcrowded and spoils the peace and quiet. Visitors' dogs can foul the beaches and footpaths which makes it unpleasant for sunbathers and walkers.

The solutions adopted to deal with the identified land use conflicts

An **Aquatic Management Plan** encourages quiet areas. **Zoning** of areas at Poole Harbour ensures that different activities are kept apart. Speed limits have been put in place to minimise the risk of accidents in the sea. World Heritage Site status allows local authorities to safeguard the coastline from over-development with strict planning controls. Public education schemes are designed to promote **responsible tourism** through guide books, leaflets and signs. Local authorities comply with the European Union's **Blue Flag Scheme** on beach quality. This helps to ensure high-quality beaches and minimal levels of sea pollution. Designated bins have been positioned in popular walking areas to stop dogs' dirt being left on beaches and paths. Signs are placed on some beaches forbidding dogs from going there.

☄ Make the Link

Unlike the Cairngorms, the Dorset coastline is not protected by National Park status. Many different organisations are responsible for its management and protection.

🔵GO! Activity 2: Paired activity

1. In pairs, design and make a **story board** to highlight a coastal land use conflict.
2. Your story board should include:
   - An eye-catching title.
   - Colourful pictures, photos and diagrams showing why there is conflict between two different land users on the Dorset coast.

Try to include some solutions to the conflict in your illustrations too.

Activity 3: (National 5)

Individually, write a **blog** which describes the importance of the Dorset coast and the strategies which help to preserve it. You should include:

- A map showing the location of the Dorset coast.
- The reasons why it was designated as a World Heritage Site.
- An outline of the strategies which help to protect the coastline. You should include both management strategies such as the Aquatic Management Plan and protection strategies such as sea walls and groynes.
- A list of ways that tourists can help to protect the area when they visit.

 Make the Link

In the Tourism chapter, you will learn the impact of tourism and ways to manage it in different contexts.

Summary

In this chapter you have learned:

- The four processes involved in coastal erosion.
- The formation of cliffs, caves and arches, stacks, headlands and bays, spits and sand bars.
- Land uses appropriate to coastal landscapes including farming, forestry, industry, recreation and tourism, water storage and supply, and renewable energy.
- Some land use conflicts within coastal landscapes and the solutions adopted to deal with these land use conflicts.

You should have developed your skills and be able to:

- Locate on a map named examples of different coastal landscapes in the UK.
- Identify and describe coastal landscape features on an OS map including: cliffs, caves and arches, stacks, headlands and bays, spits and sand bars.

End of chapter questions

NATIONAL 4 QUESTIONS

(a) Describe the location of coastal landscapes.

(b) What is the difference between constructive waves and destructive waves?

(c) What **three** things determine the size of the waves?

(d) Name the **four** processes of coastal erosion.

(e) What types of rocks form (a) headlands and (b) bays? **Choose from**: hard rock and soft rock.

(f) Name the features numbered 1 to 5 on the diagram below. **Choose from**: stack, cave, stump, blowhole and arch.

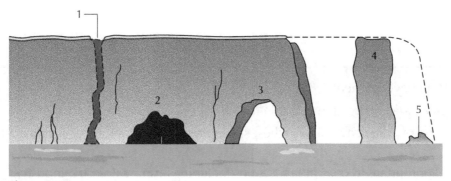

Figure 6.16

(g) What is a sand spit?

(h) Apart from tourism, name **three** different land uses on the Dorset coast.

(i) Outline a land use conflict on the Dorset coast.

(j) Suggest strategies that can reduce the conflict described above.

NATIONAL 5 QUESTIONS

(a) **Describe** the formation of a wave-cut platform.

(b) **Explain** how headlands and bays are formed.

(c) Using diagrams, **describe** the formation of sea caves, arches and stacks.

(d) **Explain** the process of longshore drift.

(e) **Describe** the formation of a sand spit.

(f) **Outline three** common land uses of a coastal landscape.

(g) **Discuss** the conflicts between land users on the Dorset coast.

(h) **Describe, in detail** an example of a **land use conflict** on the Dorset coast.

(i) **Outline** the solutions adopted to deal with the land use conflicts.

(j) **Describe** the role of an organisation that is involved in managing the Dorset coastline.

NATIONAL 5 EXAM-STYLE QUESTIONS

**You can find sample answers to these exam-style questions on the Leckie website:
https://collins.co.uk/pages/scottish-curriculum-free-resources**

1   **Explain** the formation of headlands and bays.

You may use diagram(s) in your answer.

(4 marks)

2

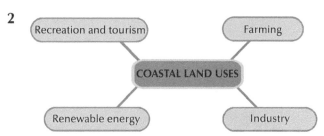

Diagram Q2: *Land uses in a coastal area*

Study Diagram Q2

Describe, in detail, land use conflicts that occur in a coastal area that you have studied.

(5 marks)

3   **Describe** the solutions to the land use conflicts identified in your answer above.

(5 marks)

4

Diagram Q4: *OS map extract of a coastal landscape, scale 1:50 000*

Look at the OS map above.

Describe different coastal features located on the map extract.

You must include features of erosion **and** deposition in your answer.

(4 marks)

LEARNING CHECKLIST

Now that you have finished the **Coastal landscapes** chapter, complete a self-evaluation of your knowledge and skills to assess what you have understood. Use traffic lights to help you make up a revision plan to help you improve in the areas you identified as red or amber.

- Locate on a map named examples of different coastal areas in the UK.

- Label the different features on a diagram of a coastal landscape.

- List the names and meanings of key coastal features.

- Explain the processes of:

  ➢ hydraulic action

  ➢ abrasion

  ➢ solution

  ➢ attrition

- Describe the formation of a wave-cut platform.

- Explain how headlands and bays are formed.

- Describe the formation of sea caves, arches and stacks.

- Explain the process of longshore drift.

- Describe the formation of a sand spit.

- Describe the formation of a sand bar.

- Identify and describe coastal landscape features on an OS map including: cliffs, caves and arches, stacks, headlands and bays, spits and sand bars.

- Outline the use of a coastal landscape.

- List the conflicts between land users on the Dorset coast.

- Outline solutions adopted to deal with the identified land use conflicts.

Glossary

Abrasion: When rocks carried by the waves wear away the coastline as they are thrown against headlands by the force of the waves (also called corrasion).

Alternating bands: Rocks which are laid down in different sections.

Aquatic Management Plan: A report which outlines the safe and responsible use of the sea along the Dorset Coast.

Arable farming: When farmers grow crops on their land.

Arch: A natural rock formation often created by two sea caves eroding backwards towards each other until the back walls disappear.

Attrition: When rocks and pebbles carried by the waves smash into each other, wearing each other away and gradually becoming smaller, rounder and smoother.

Backwash: The movement of beach material from land to the sea.

Bay: A low-lying inlet of land on the coast.

Beach: A coastal landform which usually contains sand, shingle or pebbles.

Bird watching: An activity which involves people observing different species of birds.

Blowhole: The hole at the top of a headland.

Blue Flag Scheme: A scheme used to ensure safe, clean beaches and seas for tourists.

Cave: A small opening in a headland.

Clay: A type of sedimentary rock.

Cliff: A vertical rock face.

Coastal footpaths: A trail that follows the line of the coast.

Coastal landscape: A particular type of landform located where the land meets the sea.

Coastal scenery: Natural landforms that are located along the shoreline, where the land meets the sea.

Coastline: The area where the land meets the sea.

Conserve: To protect an area from damage or development to maintain its natural beauty and existence.

Constructive waves: Waves that build up material on the coastline.

Corrasion: When rocks carried by the waves wear away the coastline as they are thrown against headlands by the force of the waves (also called abrasion).

Corrosion: When chemicals in the seawater dissolve minerals in the rocks, causing them to break up (also called solution).

Curved hook: The end of a sand spit is often shaped in this way by a change in wind direction which makes it curve into a hook.

Deposition: The dumping of rocks at a location.

Depositional feature: A landform which has been formed through the dumping of beach material.

Destructive waves: Waves that remove material from the coastline.

Differential erosion: When different types of rocks are eroded at different rates.

Dissolve: When something is broken up in a liquid and no longer exists.

Erosion: The wearing away of the landscape.

Estuary: The area where river water meets sea water when a river enters the sea.

Farming: The growth of plants or rearing of animals for food.

Fetch: The distance the wind blows over the sea.

Forestry: When an area of natural woodland or trees is planted by humans.

Fossil hunting: An activity which involves people searching for imprints of ancient plants or animals in rocks.

Geology: The study of different rock types.

Gravity: A natural force which contributes to the movement of beach material.

Harbour: A sheltered area of water that boats and ships use.

Headland: A high area of land that extends out into the sea.

Honeypot: An area which has many tourist facilities to attract visitors and keep them away from fragile areas.

Hydraulic action: When the power of the waves forces air into cracks in rocks, compresses it and blows the rock apart as the pressure is released.

Industry: The type of work that people do or business activity.

Lagoon: A shallow area of water separated from the sea by a sand bar.

Land use conflict: A disagreement over the way an area of land is used.

Limestone: A type of sedimentary rock.

Line of weakness: A part of a rock that is more susceptible to erosion.

Litter picks: When groups of people arrange to clean up an area by lifting rubbish and carefully disposing of it.

Longshore drift: The process by which material is moved along a coastline.

Marram grass: A long type of grass found growing on coastal sand dunes.

Mussel farming: When mussels (a type of seafood) are raised for human consumption.

National Trust: A charity responsible for ensuring the protection and preservation of historic places and spaces for public enjoyment.

Nature reserve: A protected area where plants and animals are conserved.

Nuclear power station: A thermal power plant where the heat source is a nuclear reactor.

Oil field: An area which has many oil wells for extracting petroleum from underneath the land or sea.

Oyster farming: When oysters (a type of seafood) are raised for human consumption.

Pastoral farming: When animals are reared, e.g. beef cattle, for their meat.

Point: The tip of a headland.

Prevailing wind: The most common wind direction.

Quarrying: The removal of rocks such as slate and granite from the landscape.

Recreation: An activity undertaken during leisure time.

Renewable energy: Energy which is generated from sources that are continually replenished e.g. wind, wave and solar power.

Reservoir: A man-made body of water used to store drinking water.

Resistant rock: Hard rocks which are less easily eroded by the sea.

Responsible tourism: When visitors are encouraged to cause minimal damage to an area.

Salt marsh: A coastal ecosystem located between land and open salt water that is often flooded by the sea.

Sand bar: A linear landform extending into the sea and across a bay.

Sand dunes: Small ridges or hills of sand found at the top of a beach.

Sand spit: A depositional feature connected to the coastline (also called a spit).

Shoreline: Where the land meets the sea.

Site of Special Scientific Interest: A designated area in the UK which is protected.

Solution: When chemicals in the seawater dissolve minerals in the rocks, causing them to break up (also called corrosion).

Souvenir shop: A retail outlet that sells tourist goods, e.g. postcards and keyrings.

Spit: A depositional feature connected to the coastline (also called a sand spit).

Stack: A tall piece of rock separated from the headland.

Stump: A small piece of rock separated from the headland.

Swash: The movement of beach material from the sea to the land.

Tourism: Travel for recreation or business purposes.

Traffic congestion: When a large volume of vehicles cause traffic jams.

Transportation: The movement of rock from one location to another.

Undercutting: The wearing away of the base of a cliff.

Water skiing: A type of water sport where a person wearing skies is pulled behind a boat.

Wave-cut notch: When the sea erodes the base of a cliff creating a hollow space.

Wave-cut platform: The flat area of rock located at the base of a cliff which represents the remnants of the headland that has been eroded by the waves.

Waves: The movement of water when affected by the wind.

Wildlife: Non-domesticated/wild plants and animals.

Wildlife habitats: The place where wild plants and animals live.

World Heritage Site status: A designated area by the United Nations which is protected.

Yachting: A type of water sport which involves sailing or boating.

Zoning: When areas are designated for different activities, e.g. swimming is in a separate zone from jet skiing.

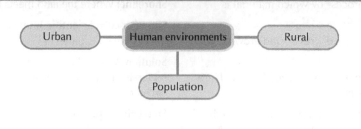

Urban — Human environments — Rural

Population

❖ Human Geography is the study of people and places, especially the impact of humans and their activities on the environment. By looking at different urban areas throughout the world we compare a typical city in Scotland with those in other nations.

❖ Studying different rural environments enables us to determine the impact of modern developments in farming on the natural landscape. Changes in farming throughout the world have occurred to keep pace global population growth.

❖ By analysing both physical and human factors, we can learn about world population distribution. Studying developed and developing countries, makes it possible to compare the development of countries throughout the world.

Consider the different towns and cities that you have visited and their different land uses.

Reflect on nearby farms and non-farming activities that they have e.g. farm shop.

Think about the number of children your parents had, in comparison to your great-grand parents.

The Human Environments

7 Population

Developed and developing countries

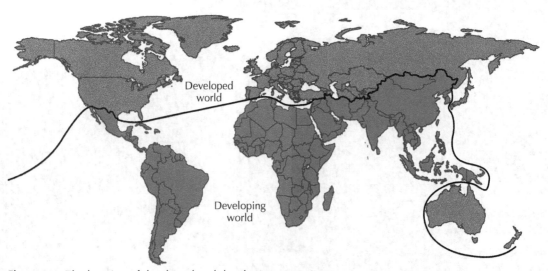

Developed world

Developing world

Figure 7.1: *The location of developed and developing countries*

Development is when there is an increase in the **standard of living** of people in a country. Countries with a high standard of living are classed as **developed** for many reasons such as sophisticated **education** systems, advanced health care and a variety of industries. Countries with a low standard of living are classed as **developing** because they have only basic schooling, few doctors and many people working in **subsistence farming**. To determine if a country is developed or developing, we need to look at indicators of development. These include economic aspects such as **wealth** and social aspects such as **health**. To analyse the development of a country, both **social and economic indicators** can be used.

Social and economic indicators

Social indicators

Social indicators measure the quality of life in a country and include:

Aspect of standard of living	Indicator
Health	Number of people per doctor.
Life Expectancy	Average age people are expected to live to at birth.
Education	Percentage of adults who are **literate**.
Food	Number of **calories** consumed per person per day.
Death Rate	Number of deaths per 1000 people per year.
Birth Rate	Number of babies born per 1000 women per year.
Infant Mortality Rate	Number of babies born per 1000 who die before the age of one.

Economic Indicators

Economic indicators measure the economic output or wealth of a country and include:

1. **Gross National Product** (GNP) per person which is the value of goods produced in a country divided by the total population.
2. **Average income per person** per year (in US dollars).
3. **Gross Domestic Product** (GDP) per person which is the value of goods and services in a country divided by the total population.

Problems with using ONE indicator

Problems linked to using one indicator to measure the development of a country are:

- Average figures are unreliable if the population has been incorrectly counted.
- The wealth of a country does not take into account how the money is used – it could be misspent by a corrupt government.
- GNP is an average, therefore the total can be easily skewed by a few very wealthy families, e.g. in Kuwait, a few wealthy oil barons hold the majority of the wealth.
- GDP does not take into consideration the cost of goods in a country as this affects what people can buy with their wages.

 HINT

To get a more accurate account of the level of development in a country, it is better to look at both social and economic indicators.

- Indicators of development are averages so they do not take into account differences between **urban** and **rural areas**, or, for example, the differences between poor **favela** areas and richer **inner city** areas in Rio de Janeiro in Brazil.

- One indicator does not give a true picture of development in a country. It is much better to examine both social and economic indicators.

Physical and human factors influencing global population distribution

HINT

Consider the reasons why India is densely populated and Greenland is sparsely populated.

Look at the map below. It shows the distribution of the world's population. Notice how the population density varies throughout the world: some places like India have a **high population density** while others are almost empty and have a very **low population density**, e.g. Greenland.

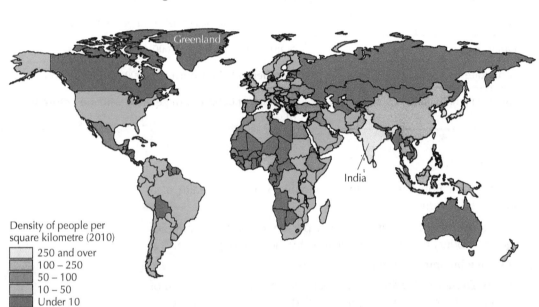

Density of people per square kilometre (2010)

- 250 and over
- 100 – 250
- 50 – 100
- 10 – 50
- Under 10

Figure 7.2: *Global population distribution*

Population density and distribution

Population density is the number of people living in an area. It is usually measured by the number of people living per square kilometre. Crowded areas have a high population density and are said to be **densely populated**. Other areas have a low population density and are said to be **sparsely populated**. Scotland has a medium population density overall because some parts, such as the city of Glasgow, have a high population density while other areas like the Highlands have a low population density.

HINT

Developed countries have a lower population density than developing countries.

The table below shows the average population density for different continents. The most densely populated continent is Asia with 93·5 people per km² and the least populated continent is Australasia with 4·5 people per km². Population density also varies within individual countries, e.g. the Amazon rainforest in Brazil has very few people living there while large cities like Rio de Janeiro in south-east Brazil have millions of people living there.

Continent	Total population	Total area (km²)	Average population density (people per km²).
Africa	1 033 000 000	30 065 000	34·4
Asia	4 167 000 000	44 579 000	93·5
Australasia	35 000 000	7 687 000	4·5
Europe	733 000 000	9 938 000	16·4
North America	352 000 000	24 256 000	14·5
South America	598 000 000	17 819 000	33·5

Population distribution is the location of people across the world, i.e. **where people live**. The population of the world is spread unevenly across the globe with large numbers of people living in the same area. Overall, the world has more empty areas than crowded areas, e.g. according to the United Nations, more people now live in towns and cities than in rural areas. There are several **physical** and **human factors** which help to explain the distribution of world population.

Figure 7.3: *In cities, people live very close together*

Figure 7.4: *In other parts of the world, people are more scattered across the landscape*

HINT

Distribution is **whereabouts** people are located, e.g. along a coastline or in towns and cities.

Physical factors influencing global population distribution

Climate

People prefer to live in **temperate climates** where there is enough rainfall and no temperature extremes, i.e. not too hot or too cold.

Few people tend to live in areas with extreme climates for a number of reasons:

CLIMATE TOO HOT e.g. Sahara desert (Figure 7.5)	CLIMATE TOO COLD e.g. Arctic Canada or mountainous regions (Figure 7.6)	CLIMATE TOO HUMID e.g. Amazon rainforest (Figure 7.7)	CLIMATE TOO DRY e.g. hot and cold deserts (Figure 7.8)
Very high temperatures make farming difficult as animals need grass and water.	Very cold temperatures make it hard to grow crops. Food has to be imported which is expensive.	Rainforests are uncomfortable to live in due to high temperatures and high rainfall.	A lack of water prevents crop growth.
A lack of drinking water prevents people living in areas.	**Frostbite** makes it difficult to work outside and protective clothing needs to be worn.	Diseases like malaria spread easily.	**Dehydration** is a problem in hot and dry areas.
Sunburn and **heat stroke** are health issues caused by intense heat.	**Permafrost** makes building houses and roads difficult as the ground is frozen for much of the year.	Thick vegetation growing in the climate makes building difficult.	Soil dries out and turns to dust, so difficult to grow crops.
It is impossible to grow crops without **irrigation**.	Very cold weather means high heating costs.	Soil quality is poor once the canopy of trees is removed, so it is difficult to grow food.	Irrigation is very expensive.

Figure 7.5

Figure 7.6

Figure 7.8

Figure 7.7

Relief

People favour living on flat, low-lying areas as they are easier to build on and grow crops. Coastal areas allow trade to take place as ports locate by the sea. **Tourism** is also popular on coastal areas. This means a variety of jobs are available, attracting large populations.

Few people tend to live in mountainous areas because:

Steep slopes make it difficult for machinery to operate. It is also difficult to build houses, factories and **transport links**. Upland areas are too cold and wet which makes it difficult to grow crops. Mountainous areas are often **isolated** which makes them difficult to access. This deters industries from locating in these areas.

Soils

People prefer to live in areas with fertile soils so that crops can be grown to supply food. Where there are poor-quality soils, e.g. on steep slopes, few crops can be grown so less people live there.

Natural resources

People tend to live in areas where there are minerals such as gold and raw materials like timber to exploit and sell. Natural landscapes with beautiful **scenery**, such as white, sandy beaches in Thailand and snow-capped mountains in the Swiss Alps, are a natural resource which attracts tourists. Tourism generates job opportunities in hotels, shops and restaurants. People live where they can benefit from tourist-related jobs, e.g. coastal Spain.

People tend not to live in areas with few natural resources because few natural resources means there will be a lack of **industry** and this in turn means less **employment opportunities** for people who may live there.

Availability of water

People are likely to live in areas where there is a supply of drinking water. Water is needed for human survival and without it crops won't grow and animals cannot survive. People cannot live without water so tend to settle where there is a fresh supply.

 HINT

Consider the impact of tourism on the economy of a country with beautiful beaches.

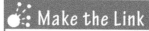 Make the Link

In the Tourism chapter you will learn about the benefits and problems created by tourists.

Human factors influencing global population distribution

Job opportunities

Jobs in different industries encourage people to move to find work.

Transport and communications

Areas which are accessible tend to have higher population densities. Places with good transport links such as roads, railways and airports attract people and industries which in turn creates employment opportunities. In contrast, remote areas which are isolated and have poor transport links do not attract people or industries.

Government aid

Industries locate where there is government funding available. In turn, people move into these areas for work.

Services

Many urban areas (towns and cities) are crowded as people move to cities like Paris, London and New York for a variety of amenities and services, e.g. education, health care facilities, a variety of jobs and entertainments.

Technology

Advances in farming technology, e.g. irrigation and fertilisers, allow people to move to areas and farm where it had previously not been possible. Developments in construction, e.g. bridges and tunnels, allow areas to be accessed and used that previously would not have been accessible.

Migration

People moving from one place to another can change the population distribution in different places. Refugees can crowd into camps located in nearby countries to get away from civil war in their own country, e.g. Rwandans moving to Tanzania.

World population growth

Figure 7.9 shows that the world's population has and will continue to increase. Population growth was fairly low and steady until the 1950s when there was a population 'explosion'. Since then, the world's population has rapidly increased and in 2011 it passed 7 billion people.

Make the Link

In the Urban chapter, you will learn why different industries locate in certain areas.

Make the Link

Think about how **physical landscapes** affect population distribution in different places.

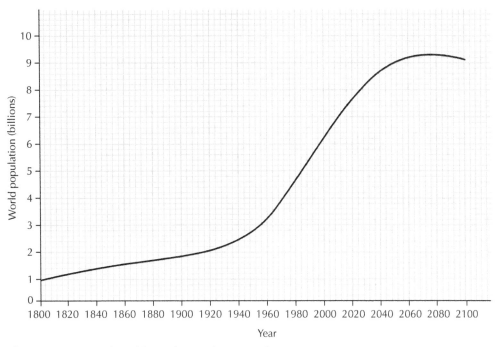

Figure 7.9: *Projected world population change until 2100*

Population growth in developed and developing countries

The majority of world population growth is taking place in developing countries. This is shown in Figure 7.10 below:

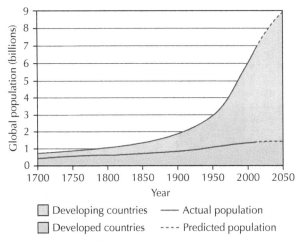

Developing countries — Actual population
Developed countries ---- Predicted population

Figure 7.10: *Population growth compared in developed and developing countries*

Within developing countries, the population of urban areas is growing rapidly as people move from rural areas. This movement is the result of both 'push' and 'pull' factors. These are shown in Figure 7.11.

149

PUSH FACTORS	PULL FACTORS
• Lack of running water	• Schools available for children's education
• Lack of basic services such as electricity	• Lots of shops to buy food
• Long distances to schools and hospitals	• Running water and electricity
• Farming is hard without machinery	• Plenty of entertainment and things to do
• Wealthy landowners buy machines and local workers are made unemployed	• A wide variety of jobs
• Population growth means food shortages and less land to farm	• Doctors, medicines and hospitals
• Natural disasters (flood, famine etc.) destroy crops	• The possibility of a better quality of life

Figure 7.11: *Reasons for the growth in urban areas in developing countries*

Population structure of developed and developing countries

Look at the population pyramids shown in Figure 7.12 and 7.13. They show the structure of the population in a developed country (UK) and developing country (India).

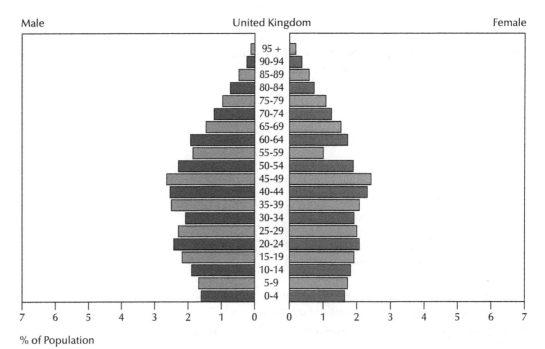

% of Population

Figure 7.12: *Population structure of a developed country: UK*

The population structure for a developed country like the UK is as follows:
- Low birth rate
- Low death rate
- Low infant mortality rate
- High life expectancy.

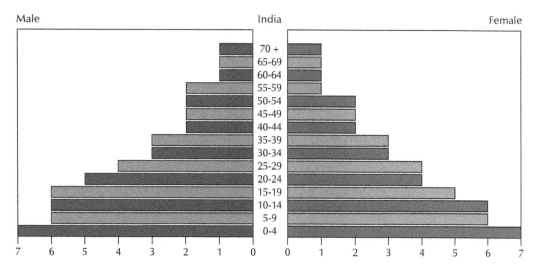

% of Population

Figure 7.13: *Population structure of a developing country: India*

The population structure for a developing country like India is as follows:

- High birth rate
- High death rate
- High infant mortality rate
- Low life expectancy.

Factors affecting birth and death rates

Factors affecting birth and death rates in developed countries like the UK

The population in the developed world is fairly **stable** and neither rising nor falling considerably. In **developed countries birth rates are low** for a number of reasons: women's status in society has improved and they are no longer seen as solely child bearers. For many modern 21st century women, careers are more important than having a family. People marry later so there is less opportunity to have large families. The cost of raising a family has increased and couples cannot afford to have many children. There is now an increased desire for material possessions, e.g. expensive cars and exotic holidays instead of having children. **Contraception** is more readily available and free on the NHS in the UK. **Family planning** clinics enable women to seek advice on bearing children.

In **developed countries, death rates are low** for a number of reasons: advances in modern medicine such as life support machines keep people alive. Children and elderly people are given vaccinations to prevent **diseases**, e.g. measles and flu. Health care is free in the UK and advances in medical science means that more people are kept alive through life-saving operations such as heart surgery. Improvements

🔍 HINT

A population pyramid is a double bar graph – turn the graph on its side to see how the numbers of males and females decrease towards the top of the graph.

151

Make the Link

In the Rural chapter, you will study the impact of agricultural changes on rural areas.

in clean piped water and **sanitation** ensure that other diseases like typhoid and cholera are prevented. Modern technology such as fertilisers and irrigation enhanced food production so shortages have been eliminated. Food storage and refrigeration have improved food quality and quantity. Increased **world trade** means that developed countries can afford to import a variety of food stuffs from around the world to ensure that a balanced **diet** is available. Lower **infant mortality rates** means that people do not need to have as many children to ensure that some survive.

Factors affecting birth and death rates in developing countries like India

In **developing countries** like India, death rates have also decreased but birth rates remain high. This means that there is a **population growth** in developing countries. Birth rates are high for a number of reasons: children are needed to work and bring in an income for the family. They are also required to look after parents in old age due to a lack of **pensions**.

More children are born as there is an absence of contraception and knowledge of how to use it as literacy rates are often low. Family planning clinics are often unavailable as there is a lack of health care services. Parents have more children in the hope that some survive as infant mortality rates are high. Certain religions such as Catholicism encourage large families. In some areas children are viewed as a sign of virility and so people have more babies.

In **developing countries death rates are high** for a number of reasons: a lack of crops due to **drought**, **famine** or **natural disasters** means that food supplies are uncertain. As a result, people often suffer from **malnutrition** and **starvation**. Poor sanitation and lack of clean drinking water means that people are more susceptible to diseases like Dysentery. High levels of endemic diseases like Malaria increase death rates in developing countries. Poor health services such as a lack of doctors and medicines means that people die from treatable illnesses like diarrhoea. War, e.g. in Afghanistan, increases death rates in developing countries.

Make the Link

In the Environmental hazards chapter, you will study the impact of natural disasters like earthquakes, volcanoes and tropical storms on death rates in different countries.

Population change

A country's population will increase through births and **immigration** and decrease when people die or **emigrate**. For a country's population to grow naturally, each couple must have more than two children: one to replace the mother when she dies and one to replace the father when he dies. The **fertility rate** refers to the average number of children born to a woman in her lifetime. In developed countries the fertility rate is often very low as many women have few children, whereas in many developing countries the fertility rate is high as the majority of women have many children.

Global population increase is the result of a decrease in world death rates while birth rates remain high. The difference between birth and death rates is known as the **natural increase**. In developing countries, like India, the natural increase is high. Some developed countries, like Sweden, are experiencing a **natural decrease** as the birth rate has fallen below the death rate. The **demographic transition model** is essentially a multiple line graph and shows the different stages that a country goes through towards development. Figure 7.14 shows the location of places at different stages on the demographic transition model. Figure 7.15 is the demographic transition model and illustrates how birth rates, death rates and total population change over time. India's development is currently at stage 3 of the demographic transition model while Sweden is at stage 5.

> ### ○ HINT
>
> Scotland's population is experiencing a **natural decrease** BUT overall the total population is increasing due to immigration.

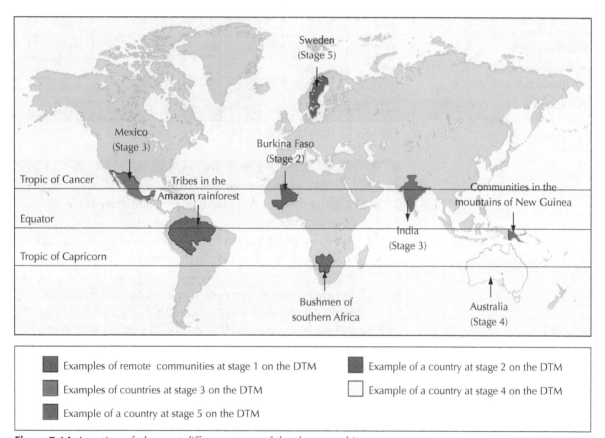

Figure 7.14: *Location of places at different stages of the demographic transition model*

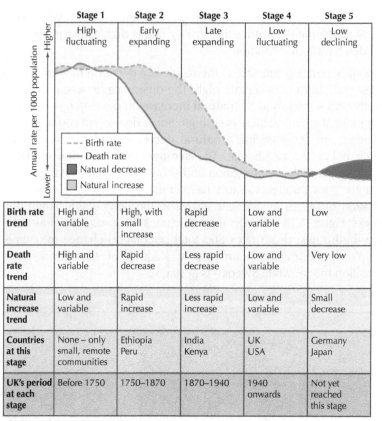

	Stage 1	Stage 2	Stage 3	Stage 4	Stage 5
	High fluctuating	Early expanding	Late expanding	Low fluctuating	Low declining
Birth rate trend	High and variable	High, with small increase	Rapid decrease	Low and variable	Low
Death rate trend	High and variable	Rapid decrease	Less rapid decrease	Low and variable	Very low
Natural increase trend	Low and variable	Rapid increase	Less rapid increase	Low and variable	Small decrease
Countries at this stage	None – only small, remote communities	Ethiopia Peru	India Kenya	UK USA	Germany Japan
UK's period at each stage	Before 1750	1750–1870	1870–1940	1940 onwards	Not yet reached this stage

Figure 7.15: *Demographic transition model*

HINT

- Demographic means it is something to do with populations.
- Transition means that change is taking place.
- A model is a way of showing what often happens in real-life.

GO! Activity 1: Paired activity

1. In pairs, collect small pieces of card and make up a population game.
2. Draw a population pyramid on the front of each card and write a description of the population structure on the back of it.
3. Test each other to see if you can describe the population without using your book or notes in your jotter! You must include birth rates, death rates and life expectancy in your descriptions.

GO! Activity 2: (National 5)

Individually, undertake computer-based research and write a report detailing why:

1. More people live in urban areas than rural areas in developed countries.
2. Many people are moving from rural areas to urban areas in developing countries.

Your report must include:

- A title
- A definition of an urban area and a rural area.
- At least six different reasons explaining why more people live in urban areas than rural areas in developed countries.
- At least six different reasons why many people are moving from rural areas to urban areas in developing countries.
- Pictures and graphs to illustrate the text.

Summary

In this chapter you have learned:

- The difference between developed and developing countries, and their location.
- Social and economic indicators and how they demonstrate the level of development in a country.
- Physical and human factors influencing global population distribution.
- Factors affecting birth and death rates.

You should have developed your skills and be able to:

- Interpret population distribution on a map(s).
- Analyse population structure on a population pyramid(s).
- Examine population change on the demographic transition model.
- Use research skills to collect and interpret geographical information.

End of chapter questions

National 4 questions

(a) Describe the location of developed **and** developing countries.

(b) Do developed countries have a high **or** low standard of living?

(c) Which of the following is an **economic** indicator of development? **Choose from**: birth rate; average income per person; life expectancy.

(d) What is the difference between **population density** and **population distribution**?

(e) Copy and complete the table below:

Human factors affecting population distribution.	Physical factors affecting population distribution.

(f) Is the world's population increasing **or** decreasing?

(g) Describe the birth rates **and** death rates in a **developed** country.

(h) Are the birth rates **and** death rates in a **developing** country high **or** low?

(i) Describe the life expectancy in a **developing** country.

(j) Explain why death rates are **higher** in developing than developed countries.

National 5 questions

(a) Choose **one social indicator** of development and **describe** how it shows the level of development in a country.

(b) Describe **three problems** with using only one indicator to measure the development of a country.

(c) **Give reasons** why people **do not** locate to mountainous areas.

(d) **Explain** why people choose to live in **urban** areas.

(e) Look at Figure 7.12. **Describe in detail** the structure of the UK's population. You must refer to birth rates, death rates and life expectancy.

(f) **Compare** the differences between the population structure of the UK and India.

(g) **Explain** why birth rates are low in developed countries like the UK.

(h) Look at Figure 7.15. Sweden is at stage 5 on the demographic transition model. **Describe** the population of Sweden. You must include birth rates, death rates and total population.

(i) **Give reasons** to explain why Sweden has a **natural decrease**.

National 5 exam-style questions

You can find sample answers to these exam-style questions on the Leckie website: https://collins.co.uk/pages/scottish-curriculum-free-resources

1

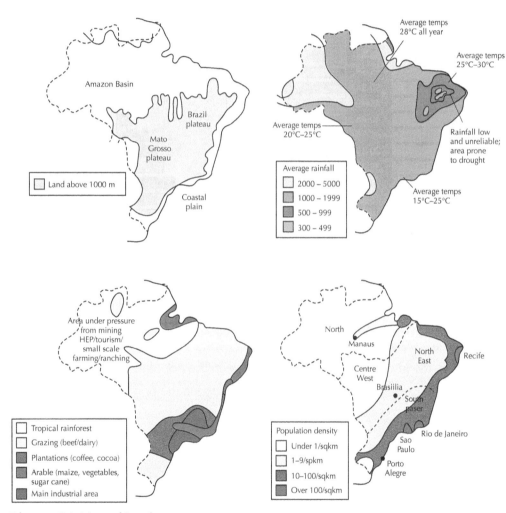

Diagram Q1: *Maps of Brazil.*

Study Diagram Q1 above and **give reasons** for the population distribution in Brazil.

(6 marks)

2 A typical exam question will ask you to describe the population structure of a developed or developing country. You may also be asked to compare the populations as shown in the sample question below.

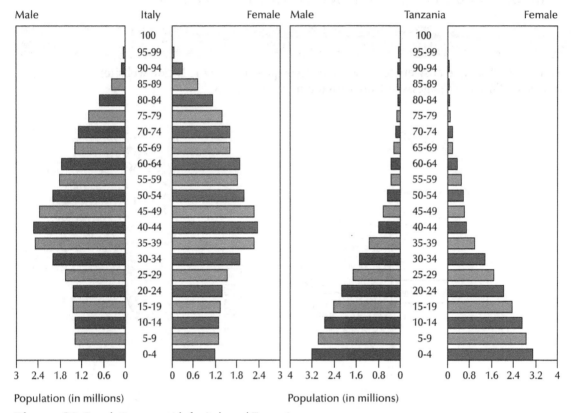

Diagram Q2: *Population pyramids for Italy and Tanzania.*

Study Diagram Q2 above.

    a) Describe in detail the differences between the populations of Italy and Tanzania.

    b) Give reasons for the differences shown between the populations of Italy and Tanzania.

(5 marks)

3

Year	Birth rate per 1000 people	Death rate per 1000 people
1970	38	16
1990	32	10
2007	23	8

Birth rates and death rates for India

Study the table above.

Explain ways in which developing countries such as India are reducing birth rates.

(4 marks)

LEARNING CHECKLIST

Now that you have finished the **Population** chapter, complete a self-evaluation of your knowledge and skills to assess what you have understood. Use traffic lights to help you make up a revision plan to help you improve in the areas you identified as red or amber.

- Describe the terms developed country and developing country.

- Give two examples of social indicators.

- Explain how social indicators demonstrate the level of development in a country.

- Give two examples of economic indicators.

- Explain how economic indicators demonstrate the level of development in a country.

- List the physical factors that influence global population distribution.

- Explain how physical factors influence global population distribution.

- List the human factors that influence global population distribution.

- Explain how human factors influence global population distribution.

- Outline the factors affecting birth rates in developed countries.

- Outline the factors affecting birth rates in developing countries.

- Outline the factors affecting death rates in developed countries.

- Outline the factors affecting death rates in developing countries.

- Interpret maps and explain the links between population distribution and physical and human factors.

- Analyse population structure on a population pyramid(s).

- Examine population change on the demographic transition model.

- Research a country and display the information on two different graphs.

Glossary

Birth rate: The number of babies born per 1000 women per year.

Calories: The units of energy contained in food and drink.

Climate: The usual weather that a place experiences over an average period of 30–40 years.

Contraception: Methods used to prevent pregnancy, e.g. condoms.

Death rate: The number of people who die per 1000 per year.

Dehydration: When the normal water content of a human body is reduced.

Demographic transition model: A multiple line graph showing how birth rates, death rates and total population change as a country moves through different stages of development.

Densely populated: An area that is crowded.

Developed country: A country where people have a high standard of living.

Developing country: A country where people have a low standard of living.

Development: The process of becoming an advanced country in terms of wealth, health and technology.

Diet: The type and amount of food consumed by people.

Disease: An abnormal condition that affects the body, e.g. malaria.

Drought: When there is a lack of rainfall over a long period of time.

Economic indicator: A statistic about the economy that can be used to illustrate a country's level of development, e.g. GNP.

Education: A form of learning in which skills and knowledge are transferred through teaching, training and research.

Emigration: When people move from one country to another.

Employment opportunities: When there are jobs available for people to do.

Family planning: When people plan when to have children.

Famine: A widespread lack of food caused by different factors including drought (lack of rainfall).

Favela: An area of very poor-quality housing in Brazil (also called a slum or slum housing).

Fertility rate: The average number of children born to a woman in her lifetime.

Frostbite: The medical condition caused by body tissues freezing.

Global population distribution: The spread of people across the Earth's surface.

Gross Domestic Product: The value of all the goods and services produced in a country in one year.

Gross National Product: The value of all the goods produced in a country in one year.

Health: The physical and mental condition of human beings.

Heat stroke: A life-threatening condition caused by over-exposure to the sun.

High population density: A place with many people (also called densely populated).

Human factor: Something that does not occur in the natural world but is related to humans, e.g. employment opportunities.

Immigration: When people move into a country.

Industry: The type of work that people do or business activity.

Infant mortality: Deaths of children under the age of one.

Infant mortality rate: The number of children per 1000 per year who die before the age of one.

Inner city: The land use zone made up of the old industrial zone and old housing zone.

Irrigation: Artificially watering the land.

Isolated: When a place is cut-off due to lack of transport and communications.

Life expectancy: The age people are expected to live to.

Literate: Being able to read and write.

Low population density: A place with few people (also called sparsely populated).

Malnutrition: A lack of proper nutrition.

Migration: The movement of people from one place to another.

Natural decrease: When a population decreases because more people are dying than babies are being born.

Natural disaster: A devastating event, e.g. earthquake, volcanic eruption or tropical storm.

Natural increase: When a population increases because more babies are being born than people are dying.

Natural resources: Commodities that come from the environment, e.g. coal and oil.

Pension: A sum of money paid to people upon their retirement.

Permafrost: When soil is permanently frozen.

Physical factor: A naturally occurring thing, e.g. climate.

Population: The total number of people living in a country.

Population density: The number of people living in an area (Km^2).

Population distribution: The location of people in an area.

Population explosion: When the population of the world increases rapidly.

Poverty: A lack of money or material possessions.

Relief: The height and shape of the land.

Rural area: An area of countryside.

Sanitation: The treatment and proper disposal of sewage.

Scenery: The appearance of a place, especially the natural landscape.

Social indicator: A feature of society that can be measured to illustrate a country's level of development, e.g. birth rate.

Soil: A natural substance made of layers of minerals, organic matter and weathered rock.

Sparsely populated: An area that has few people.

Standard of living: How well off the people of a country are in terms of wealth, health and education.

Starvation: When people are extremely hungry due to lack of food over a long period of time.

Subsistence farming: When farmers grow only enough food for themselves and their families.

Temperate climates: The usual weather conditions in an area that does not include extremes of temperature or precipitation.

Tourism: Travel for recreation or business purposes.

Transport links: Different forms of transportation including roads, railways and ferry routes.

Urban area: A built up place, e.g. town or city.

Virility: The ability of a man to father many children.

Wealth: An abundance of material possessions or money.

World trade: The buying and selling of goods between different countries.

8 Urban

Within the context of urban areas, you should know and understand:

- Characteristics of land use zones in cities in the developed world.
- Recent developments in the CBD, inner city, rural/urban fringe in developed world cities.
- Recent developments which deal with issues in slum housing in developing world cities.

You also need to develop the following skills:

- Locate on a map named examples of developed and developing world cities.
- Identify and locate different land use zones and patterns on an OS map.
- Use research skills to collect and interpret geographical information.

Characteristics of land use zones in cities in the developed world

Land uses such as housing, shops and **industry** can be categorised into **land use zones**. Areas within towns and cities can be recognised by different land use zones as they have similar characteristics. Land use zones include the central business district (CBD), inner city (old housing and old industrial zones), new housing zone, new industrial zone, new business district and the rural/urban fringe. Figures 8.1 and 8.2 show these different land use zones in a town/city:

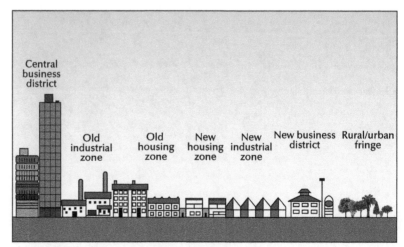

Figure 8.1: *Land use zones in an urban area*

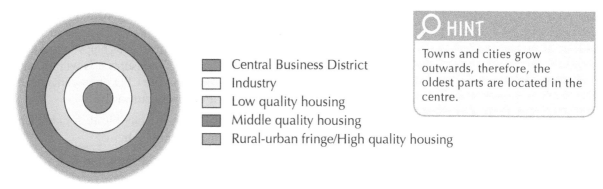

Figure 8.2: *Land use zones in an urban area: concentric zone model*

Characteristics of urban land use zones

Central Business District (CBD)

Figure 8.3: *The Central Business District/City Centre*

Figure 8.4: *The CBD of Glasgow*

The CBD is usually located at the most accessible location such as at a crossing point of a river, like Glasgow's River Clyde, and in the centre of a city where roads and railways meet. This helps to make the city an accessible location for workers and ensures it is within reach of most people for shops and businesses. The CBD often contains the oldest part of the town as urban areas grow outwards.

The CBD has the highest **land values** in the city because many land users want to locate there. It is easily identified by:

- Tall, high-density buildings and a lack of open space.
- Where main roads and railways meet.
- Main railway stations and bus station.
- Multi-storey car parks.
- Comparison shops where people can browse in the same type of shop, e.g. jewellers.
- Specialist shops, e.g. bridal boutiques.
- Large department stores and shopping centres.
- Tourist information centre.
- Cultural centres, historical buildings and museums.
- A concentration of churches and a cathedral.
- Hotels for tourists and business meetings.
- Offices, finance, banks, administration, town hall (known as the business sector).
- Entertainment facilities such as cinemas and theatres.
- High traffic and pedestrian flows at rush-hour and Saturdays, for shopping.

> ### 🔍 HINT
>
> Some of these features will be easy to pick out and help you identify the CBD on an OS map, e.g. tourist information centre, cathedral, meeting of main roads and town hall.

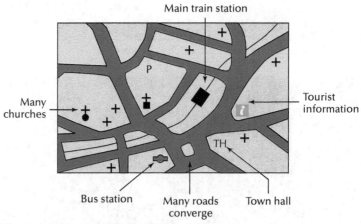

Figure 8.5: *Recognising the CBD on OS maps*

The inner city

This zone is made up of the old 19th century industrial and housing zones built during the industrial revolution. The inner city is located near the centre of the city and surrounds the CBD. Inner city areas have undergone massive regeneration in recent years due to the decline of old, **heavy industries** like coal mining, steel making and shipbuilding.

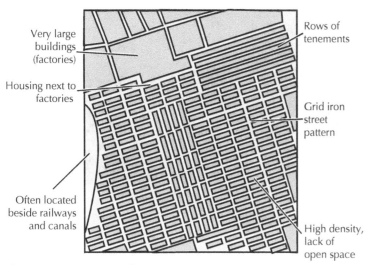

Figure 8.6: *Recognising the inner city on OS maps*

The old industrial zone

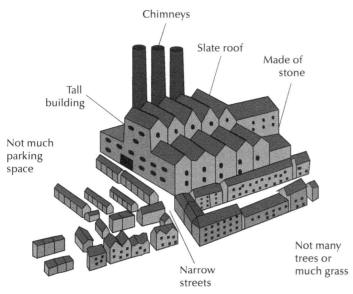

Figure 8.7: *The old industrial zone*

The old industrial zone is identified from the following features:

- Factories, warehouses, shipyards, ironworks, steel works, coal mines or spoil heaps.
- **Derelict land** – many factories closed down.
- Old industrial buildings abandoned.
- Large areas of **redevelopment** or **urban regeneration**.
- A declining population and high unemployment as older industries have closed down.
- Areas demolished and used for motorways and **ring roads**.

> 🔍 **HINT**
>
> Old industrial areas are difficult to identify on an OS map nowadays as many old factories have been demolished and the land has been redeveloped.

The old housing zone

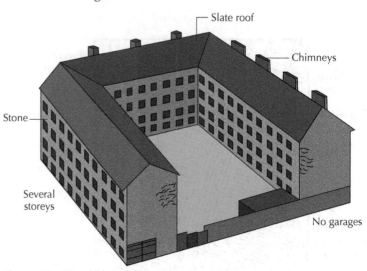

Figure 8.8: *The old housing zone*

HINT

There are no spaces for car parking or garages because cars weren't invented when these houses were built.

The old housing zone was built right beside the old industrial zone. Low-cost sandstone **tenements** were built quickly for workers in the nearby factories. People needed to live close to work as **public transport** was not developed and cars had not yet been invented. People living in poor inner city areas were able to make use of the **amenities** and **services** in the CBD.

The old housing zone is easily identified from the following features:

- Older, 19th century low-cost housing. This is most likely tenements in Scotland and **terraced housing** in England and Wales.
- **Grid iron street pattern** as housing was unplanned.
- **High density** of buildings because accommodation was needed for many workers.
- Poor-quality housing with a lack of gardens as land was precious.
- A lack of good quality open space, e.g. parks.
- No garages as there were no cars when the houses were built.
- High levels of air pollution from traffic.
- In some areas there may be visual pollution in the form of **vandalism** and **graffiti**.

HINT

When identifying the old housing zone on an OS map, remember it will be close to the CBD.

The new housing zone

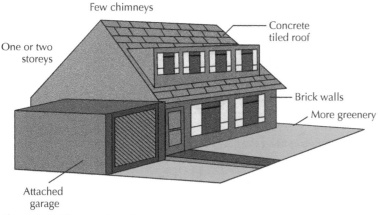

Figure 8.9: *The new housing zone*

The new housing zone is located in the **suburbs** as more land was required to build larger low-density housing with gardens and garages. As urban areas grow outwards, the most logical location for the new housing zone is on the edge of the city. A variety of house types are located on the outskirts as these areas have been designed by town planners. People can travel into the CBD for work, shopping and entertainment by car, bus or train.

The new housing zone is easily identified from the following features:

- Newer, 20th and 21st century housing. This is most likely **detached** and **semi-detached** houses as land is cheaper further away from the CBD.

- **Cul-de-sacs** and **crescent**-shaped street patterns because housing and streets were planned so they are safer for children.

- **Low density** of buildings because more space allows houses to be built outwards.

- High-quality housing with gardens due to more land availability.

- Garages for cars.

- Space to extend or build a **conservatory**.

- Low levels of air pollution because it is further away from CBD traffic and separated from industry.

- Local corner shops for everyday essentials, e.g. bread and milk.

- Local schools nearby for children.

- Quieter and attractive environment due to proximity to the countryside.

> 🔎 **HINT**
>
> When identifying the new housing zone on an OS map, look at the edge of the town/city and look for curvy street patterns and nearby schools.

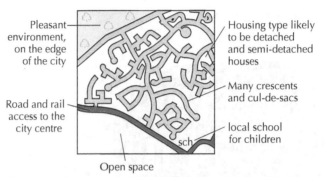

Figure 8.10: *Recognising the new housing zone on OS maps*

The new industrial zone

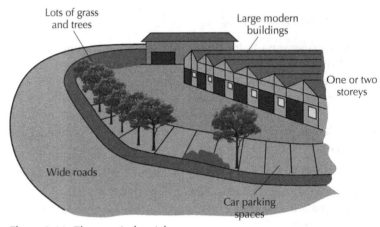

Figure 8.11: *The new industrial zone*

The new industrial zone is located further away on a **greenfield site** as more land was required to build larger low-density factories with storage space. Room is available to extend if required as land is cheaper on the outskirts. As urban areas grow outwards, the most logical location for the new industrial zone is on the edge of the city. Modern industries are located in planned **industrial estates** and are separated from housing. People can commute to work by car so industries are located beside a main road or motorway for easy access. An out-of-town location is desirable as it avoids city centre **congestion** and pollution. The new industrial zone is easily identified from the following features:

- Factories are grouped together in planned industrial estates.
- Low-density (one or two storey) buildings.
- Modern buildings built of bricks and glass.
- Wide roads for lorries to import raw materials and export finished goods.
- Separated from housing as workers commute by car.
- Landscaped with trees and shrubs to provide a pleasant working environment.
- Large car parks for workers and delivery lorries.

> ### 🔍 HINT
>
> When identifying the new industrial zone, look for large buildings beside a main road on the outskirts of the town/ city. The words 'Business Park' or 'Trading Estate' may also help.

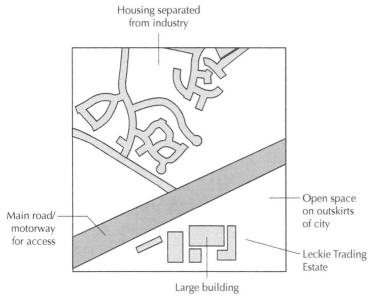

Figure 8.12: *Recognising the new industrial zone on OS maps*

The new business district

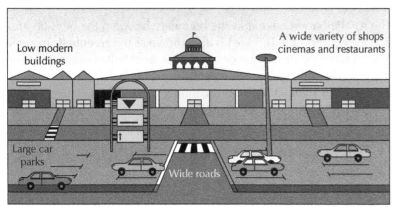

Figure 8.13: *The new business district*

The new business district is built on the edge and it is furthest away from the oldest part of the city. Large areas of land are required to build large, low-density shops, supermarkets and retail outlets. Entertainment facilities such as ten-pin bowling and cinemas are also common. Room is available to extend if required as land is cheaper on the outskirts. People can commute by car so the new business districts are located beside a main road or motorway for easy access. There are also large car parks with thousands of parking spaces for shoppers. The new business district is easily identified from the following features:

- Very large and modern low-density buildings, e.g. shopping centres, cinemas and supermarkets.
- Shops and services grouped together in **retail parks**.
- Separated from housing and industry.
- Thousands of free parking spaces.
- Beside motorway or main roads for access.

> 🔍 **HINT**
>
> When identifying the new business district on an OS map, look on the edge of the built-up area for large buildings, main roads and plenty of space.

- Built on the edge of a town with room to expand.
- Wide roads for many cars and delivery trucks.

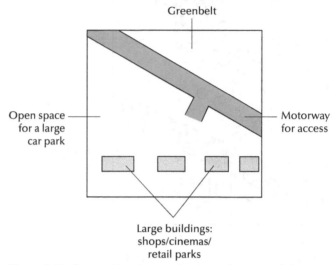

Figure 8.14: *Recognising the new business district on OS maps*

Rural/urban fringe

The rural/urban fringe marks the boundary between the built-up area and the countryside. This area is also known as the **greenbelt**. The area of land surrounding a city is designated as greenbelt to prevent the city merging with other large settlements. The greenbelt has few buildings as it is often difficult to obtain planning permission to build on this area of land.

The rural/urban fringe is easily identified from the following features:

- Open space on edge of the city.
- Farmland.
- Scattered buildings – usually farm steadings.
- Roads and railways connecting other settlements to a large city.

🔍 HINT

On an OS map remember that farmland is identified by white areas (green areas are forestry).

OS Question

Find named examples of each land use zone on the OS map of Glasgow.

🟢 Activity 1: Group activity

1. In groups, choose which city you want to study in the developed world, e.g. London would be a good example due to the many changes for the 2012 Olympics.

2. Go to the library and undertake research from books and the internet on the recent developments in the CBD, inner city and rural/urban fringe in your chosen city.

3. Individually, write a newspaper report on your city – you must show your findings in writing and on at least two different graphs.

4. Remember to include a location map of your city in your report.

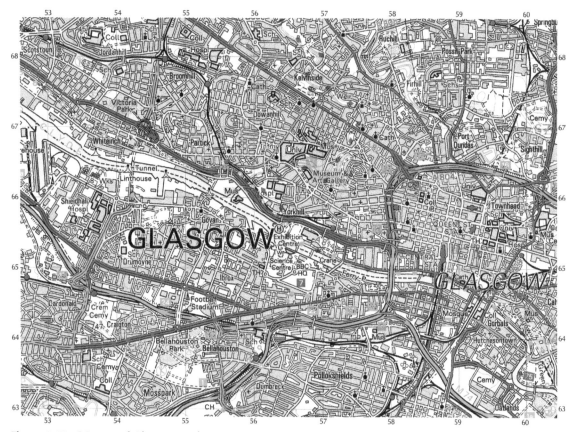

Figure 8 15: *OS map of Glasgow, scale 1:50 000*

Recent developments in the CBD, inner city and rural/urban fringe in the developed world – case study of Glasgow

There have been many developments and changes in different areas of developed world cities in recent years, including shopping, transport, industrial relocation, urban regeneration and urban sprawl. Many changes are similar throughout developed world cities but this section focuses specifically on Glasgow.

The location of Glasgow

Glasgow is Scotland's largest city by population. It is located in west central Scotland and originated on the flat flood plain of the River Clyde. The flat land was important for building and farming and the river was important as a water supply and for transportation. Glasgow's strong geographical location was also important for defence and shelter as it is surrounded by hills such as the Campsie Fells. The location of Glasgow has enabled it to grow and develop as Scotland's most popular retail centre. Various roads such as the M8, M74 and M77, and rail links, provide people with access to the city for shopping, work and entertainments.

Make the Link

In the Rivers and valleys chapter you learn about the use of the River Clyde and its valley in detail.

Figure 8.16: *Location map of the city of Glasgow*

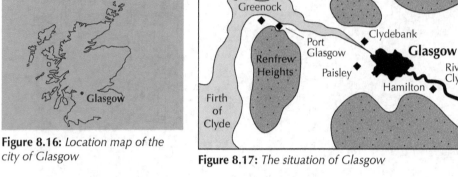

Figure 8.17: *The situation of Glasgow*

> 🔍 **HINT**
>
> Consider the impact of Internet shopping on town centre shops.

> 🔍 **HINT**
>
> Think about the shops located in your nearest town centres.

Recent developments in CBD: shopping changes

Undercover shopping centres such as Buchanan Galleries and St Enoch Centre were built to attract people into the city centre to increase revenue.

Shop closures, e.g. Borders bookshop on Buchanan Street shut down due to increased **competition** from internet shopping, **out-of-town shopping centres**, retail parks and 24-hour supermarkets.

Pedestrianised shopping streets such as Argyle Street provide a safer and less polluted shopping environment. On the other hand, cars are restricted to other areas and congest neighbouring streets.

Charity shops are now part of the 'high street' as they attempt to raise money for various causes, e.g. Oxfam, the British Heart Foundation and Cancer Research.

Discount stores, e.g. Poundland, cater for people who are less affluent.

Recent developments in CBD: transport changes

Park and ride schemes have a positive impact as there are fewer cars in the city centre which means less congestion and pollution. However public transport is becoming more expensive (particularly train fares) which discourages people from using it.

Ring roads that avoid the city centre, e.g. the M8 in Glasgow, allow through traffic to avoid CBD congestion. However, roads are still congested at peak times due to the sheer volume of rush-hour traffic.

Improved public transport, e.g. **bus lanes** help to keep buses running on time. This means that fewer people take their cars into the city centre, so there is less congestion. However, travellers are inconvenienced if services are cancelled, e.g. due to bad weather.

One-way streets such as Bath Street in Glasgow allow traffic to flow freely and move faster. On the other hand, it is difficult for drivers to get around the city centre if they are not familiar with the restrictions.

Multi-storey car parks, parking charges, double yellow lines and traffic wardens all help to reduce on-street parking. This means that fewer cars illegally park on streets and block them. However, it is expensive for drivers to use certain city centre car parks.

Congestion charges have also been introduced in London, where drivers pay a fee to take their car into the city centre.

Recent developments in CBD: the spread of the CBD into the inner city zone

Urban regeneration in some inner city areas of Glasgow, especially along the River Clyde, has enabled the CBD to grow. Other developments along the River Clyde include the Riverside Museum, BBC studios and the Hydro Music Arena. These developments have replaced old inner city industries that have declined and have enabled the CBD to expand and accommodate many leisure and tourism facilities that could not be built in the original CBD due to a lack of open space.

Figure 8.18: *Demolition of old tenement housing in Glasgow*

Recent developments in the inner city

The **inner city**, also known as the **zone of transition**, has undergone many changes due to industrial decline. The closure of many older industries left inner city areas run-down with **derelict buildings** and polluted land. High unemployment was also a huge problem. There have been a number of projects put in place to redevelop and improve inner city areas of Glasgow. This is called **urban regeneration** and has been on-going for over 50 years, when many old heavy industries started to decline and close down.

Recent developments in the inner city: the East End of Glasgow

The main focus of government funding in the East End of Glasgow, e.g. Bridgeton, involved the removal of some, and **renovation** of other, tenement housing. It also included the building of new housing, planned industrial estates and **health centres**. **Landscaping** and trees were provided to improve the run-down appearance of the areas. Private companies were involved in urban regeneration and built The Forge Shopping Centre on the site of a former steel works. Today, regeneration is on-going in the East End and one of the more recent developments is the Fort Shopping Centre. Clyde Gateway is the current regeneration program in the East End of Glasgow and in Dalmarnock, it was involved in building and converting the athletes' village for the 2014 Commonwealth Games into housing. This project has also improved roads, decontaminated polluted land and provided green spaces to the urban area.

Recent developments in the inner city: gentrification of the Gorbals

> **HINT**
>
> Contrast the appearance of old tenement housing and new modern flats in the Gorbals area.

Figure 8.19: *A new housing development in the Gorbals area*

The Gorbals area was once branded the most overcrowded **slum** area in Scotland and it had a reputation for being one of the roughest neighbourhoods in Glasgow. Urban regeneration has been extensive in some parts, and high rise flats and tenements have been demolished and replaced with new lower density housing, e.g. in Crown Street. Some land was sold to private builders who have developed luxury flats and commercial units. Landscaping has improved the appearance of some parts and local shops provide the community with services. The Gorbals continues to undergo change and it is becoming a much more pleasant and desirable place to live. Due to its proximity to the CBD, parts of the Gorbals have now been occupied by professionals who work in the city centre and want to live nearby.

Recent developments in the inner city: redevelopment of Glasgow harbour and quayside

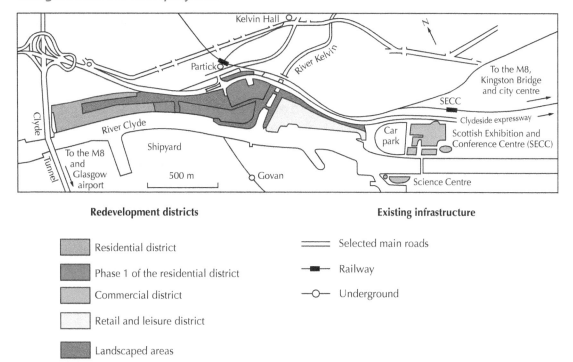

Figure 8.20: *Redevelopment along the River Clyde in Glasgow*

With the decline of old heavy industries, the city required new ways of generating income. This was done initially by building the SECC (on old **dockland** at Anderson Quay) to create income through visitors and business. New developments in recent years include the Glasgow Science Centre and IMAX Cinema which were developed on Pacific Quay. Springfield Quay has been redeveloped as a recreational outlet with cinemas, ten-pin bowling and restaurants.

Figure 8.21: *Glasgow old and new*

The River Clyde is now multi-functional for tourism, leisure, commerce and industry. Modern flats have been built with views over the Clyde to encourage people to live in the city. Transport links over the River Clyde have been improved with the addition of the Clyde Arc (Squinty Bridge) which is near the Finnieston Crane, a landmark symbolising Glasgow's engineering heritage. Landscaping and trees have also improved the appearance of the area.

Recent developments on rural/urban fringe

Urban sprawl means that countryside areas are continuously under threat from expanding towns and cities. The greenbelt is an area of countryside, usually farmland, which cannot be built on. There are strict town planning rules that normally do not permit developments to take place. The greenbelt is designed to protect rural areas from urban sprawl and prevent towns and cities from joining up. However, its protection is often a fine balance between the need for jobs and conservation. Figure 8.22 shows the developments that can take place on the rural/urban fringe if the local council grant planning permission for development.

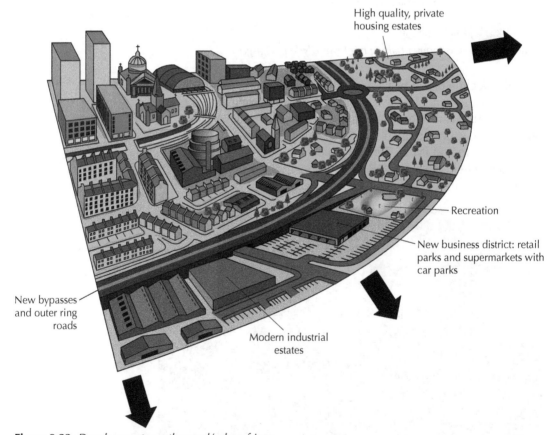

High quality, private housing estates

Recreation

New business district: retail parks and supermarkets with car parks

New bypasses and outer ring roads

Modern industrial estates

Figure 8.22: *Developments on the rural/urban fringe*

The reasons for the growth in development at the edge of the city on the rural/urban fringe are as follows:

New housing estates

The suburbs offer large amounts of unpolluted land to build on. As this area is further from the CBD, land is cheaper. It is cost-effective for developers to build outwards and there is room to build garages and gardens. In the past, **new towns** such as Cumbernauld and **council estates** like Easterhouse were built on greenfield sites to overcome the problem of overcrowding in poor-quality inner city housing such as the Gorbals. Today, new housing developments continue to add to the problem of urban sprawl as people desire to live in a quiet, out-of-town location. Small villages, e.g. Clarkston, have been eaten up by the Glasgow city boundary. Farming villages, e.g. Eaglesham, have now been turned into commuter settlements. The result is an ever expanding city boundary, loss of identity for smaller settlements and increasing house prices for local first time house buyers.

Figure 8.23: *New housing*

Modern industrial estates

Industrial estates are located on the edge of the city. Companies aim to take advantage of the cheaper land which enables the construction of large buildings and room for expansion. Good communications eg. main roads enable easy access for workers and deliveries. Nearby populations in housing estates provide a workforce and a potential **market** for the goods. However, these developments increase urban sprawl and put pressure on the greenbelt. The impact of these developments is a decrease in farmland, less greenbelt and a loss of wildlife and destruction of their habitats.

Figure 8.24: *Hillington Industrial Estate in Glasgow*

 HINT

Old industrial areas developed along the River Clyde in Glasgow to take advantage of the river as a transport route.

New business districts

Many shopping centres, retail parks and large supermarkets are located on the rural/urban fringe to take advantage of an out-of-town location that is away from city centre congestion. There is space to build large car parks for customers. People living in nearby housing estates provide a workforce and customers for their businesses. Unfortunately, these developments decrease the amount of green space and increase the volume of pollution in rural areas as more people travel by car to shop. It is worth noting that Braehead Shopping Centre, Retail Park and Intu Braehead Soar leisure facility were built on an old industrial site beside the River Clyde so it actually helped to **regenerate** a run-down area on the outskirts of Glasgow.

Figure 8.25: *intu Braehead*

Decentralisation of offices

Some offices are relocating to **business parks** on the rural/urban fringe. They are moving to take advantage of the cheaper land with room to expand. Nearby main roads provide access for workers and the unpolluted environment in the countryside helps to attract a highly skilled workforce. However, this relocation contributes to urban sprawl and **land use conflicts** with **rural land users**, e.g. farmers.

New bypasses and ring roads

Bypasses and ring roads have been built to limit the number of cars in town and city centres. This helps to reduce congestion and accidents in these busy areas. The suburbs are connected by many main roads and train lines which link to the city centre. This allows **commuters** easy access to work, shopping and entertainments. Park and ride schemes have been effective in reducing the number of cars on the roads. These schemes allow people to travel to events such as football matches without taking their cars.

Make the Link

Consider the land use conflicts that can occur on the rural/urban fringe.

> **Make the Link**
>
> In the Population chapter you learned about the differences between developed and developing countries.

> **Make the Link**
>
> In the Population chapter, you learned about 'push' and 'pull' factors that encourage people to migrate.

Recent developments which deal with issues in slum housing in developing world cities: case study of Rio de Janeiro

There have been many developments and changes in developing world cities in recent years. Many of these changes are similar but this section focuses specifically on Rio de Janeiro in Brazil.

Introduction

In **developing countries**, large numbers of people have moved from the countryside to large cities like Rio de Janeiro. This movement is known as **rural–urban migration**. People move in search of a higher **standard of living** due to rural poverty. They are attracted to cities in the hope of finding well-paid jobs in factories, better schools for their children, entertainments and better health care. However, the reality is very different and many families find themselves struggling for survival in slum housing (called **favelas** in Brazil). Roçinha is one of the largest favelas in Rio de Janeiro with a population of over 100 000 people. It consists of poor-quality, high-density housing that is badly situated on a hillside in Rio de Janeiro.

Figure 8.26: *The location of Rio de Janeiro in Brazil*

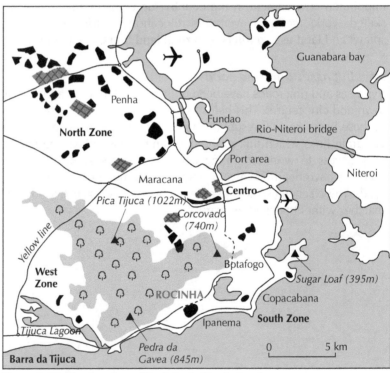

Figure 8.27: *The location of Roçinha in Rio de Janeiro*

Issues in slum housing

Poor-quality housing and overcrowding

Figure 8.28: *Housing in Roçinha*

In a favela, people build their own houses on land which does not belong to them. The land that is used is not suitable for properly built houses as it is too steep or marshy. Dwellings are made from basic materials such as wood, corrugated sheets, broken bricks and tarpaulin. The population density is very high – about 37 000 people per square kilometre. The houses lack basic amenities such as running water or toilets. Sewage often runs in open drains due to poor **sanitation** so diseases such as cholera are common and spread rapidly among the high numbers of people. As they are illegal, favelas are not maintained by the government so there is a lack of electricity, sewage

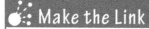

Make the Link

The average population density in South America is 33.5 people per km².

pipes, rubbish collection, schools or hospitals. There have been many unsuccessful attempts to remove favelas but as people have nowhere else to go, they usually return and rebuild their homes.

Poor health care

People live in poverty and cannot afford health care or medicines. This means that illness goes untreated and diseases spread quickly. Consequently, infant mortality rates are high and life expectancy is low at an average of 56 years in a favela in Rio de Janeiro.

Unemployment

Unemployment rates are high as there are not enough jobs to go around. Most people who do have a job work in the informal sector for 'cash in hand', e.g. cleaners. **Informal sector jobs** are very poorly paid and offer people little means of survival. Work is irregular, therefore a steady income is not guaranteed.

Crime

Figure 8.29: *Armed police in Roçinha*

Rio de Janeiro is notorious for drug trafficking, notably cocaine. Tourists visiting famous beaches such as the Copacabana are encouraged not to take expensive valuables as crime is rife. The crime rate in favelas is also extremely high as they are controlled by gangs who are involved in gun crime, drugs and murder. Roçinha is so feared by police that they do not patrol on foot without guns.

Figure 8.30: *Rubbish on street in the favela*

Pollution

Rio de Janeiro is part of Brazil's industrial triangle so air pollution from factories is a major problem. Fumes from traffic sit over Guanabara Bay and smog is common. Waste and rubbish from housing and industry create much land pollution. Seas and beaches are also polluted along the coastline. In favelas, rubbish is left to pile up and sewage runs in open drains. As a result, drinking water supplies are easily polluted. Visual pollution is caused by the slum housing themselves and can discourage tourists. Noise pollution is common due to the volume of industry, cars and people.

Landslides

Rio de Janeiro is surrounded by mountains, so, during **tropical storms, landslides** are common. Make-shift houses in favelas offer little protection to people. Houses and belongings are easily washed away by the heavy rain and mud.

Recent developments which deal with issues in slum housing

The Brazilian government knows it cannot solve the housing problem in favelas by destroying them. Instead, it wants to improve existing slum housing but, at the same time, it does not want to encourage more to develop. This is because favelas are unsightly and portray a bad image of the city. The main solutions undertaken include:

Make the Link

In the Environmental hazards chapter, you will learn about the impact of tropical storms in detail.

Self-help housing schemes: Roçinha

Make the Link

Consider the differences between the quality of housing in Glasgow and Roçinha.

Figure 8.31a: *Roçinha*

Self-help schemes are small-scale projects which let local people use their own skills to improve their housing. Favelas have a strong community spirit and people enjoy an enriched street life. As a result, the residents have managed to gradually transform the favela of Roçinha into a small city. The government works with local housing associations and has provided materials such as bricks, cement and glass to enable residents to work together to improve their own homes. In Roçinha, self-help schemes have improved the area from slums to low-quality housing. Most homes have basic facilities like electricity and others even have satellite TV. There are now services including cafes, entertainment facilities and shops run by local people. Streets have been paved and there is also some street lighting. Some people have been granted legal ownership of the land that their house is built on. However, the success of this project has been limited in certain places due to the position of many houses on steep hillsides, the sheer number of homes and the amount of funds available.

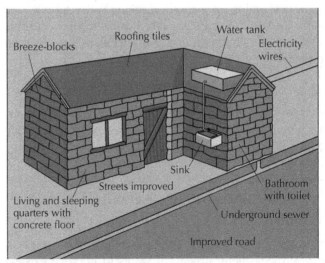

Figure 8.31b: *A site and service scheme*

Site and service schemes: the local authority Favela Bairro Project

These are **local authority programmes** designed to relocate residents from favelas. To improve 60 of the 600 favelas in Rio de Janeiro, £200 million was set aside. The initial plan was to improve the areas that were visible to tourists and people living in well-off housing areas. Brick houses were built with electricity, running water and sanitation pipes to replace dwellings made of wood or with no proper foundations. An example of such a scheme is the Favela Bairro Project (Slum to Neighbourhood project). Services in these areas include **refuse collection**, schools and health centres. Certain streets were widened to allow access for refuse collection and emergency services. Sports areas were added and streets were paved.

Within these housing areas the residents pay taxes to the government to ensure services continue to run. People can also be given permission to buy these houses.

Barra da Tijuca New Town

Figure 8.32: *Barra da Tijuca, Rio de Janeiro*

Barra da Tijuca is a **new town** built along a coastal motorway approximately 20 km from central Rio de Janeiro. Wealthier residents moved out of the overcrowded and run-down city centre. This movement is called counter-urbanisation. Housing is luxurious and spacious and residents all drive their own cars. High rise apartments make up 75% of the housing and have security entry to minimise crime. Detached houses are equipped with modern facilities. Residents are well-off and they have high paid jobs to pay for this expensive housing. Services in the town include entertainments, shopping malls and leisure centres. There are also schools, hospitals and offices. However, the housekeepers, maids, cleaners and gardeners of the wealthy people have set up their own favela in Barra!

Charities

Figure 8.33: *A charity-funded school*

Charities and aid workers help to improve the quality of life for people in slum housing. They provide money for self-help schemes and teach skills to local people. The **Developing Minds Foundation** is a voluntary organisation that builds schools and supports education programs in the favelas in Rio de Janeiro. The goal is to improve the literacy rates of children so they can get good jobs, improve their standard of living and enhance their life choices when they grow up.

Eviction and demolition programs

More extreme measures to improve Brazil's favelas in recent years have been to evict people and demolish the favelas. As Brazil was host to the 2014 FIFA World Cup and the Olympic Games in 2016, the Government wanted to improve the image of the country and attract visitors. As a result, thousands of people lost their dwelling and were forced to set up home in a new slum elsewhere.

GO! Activity 2: Paired activity

1. In pairs, design and make a mind map outlining the different land use zones in a city that you have studied.

2. Include sketches and information on each land use zone below:

   - CBD
   - Inner city
   - New housing zone
   - New industrial zone
   - New business district

3. Remember to add named examples to show that you know your case study.

GO! Activity 3 (National 5)

Imagine that you were a resident in a favela for a day. Create a diary entry outlining:

- The current issues in your home and favela.
- The recent developments which help to deal with these issues.

Your diary entry must include:

- A title
- A definition of a favela.
- At least four different problems in your favela.
- At least four different solutions that have been devised to solve these problems.
- Pictures, statistics and graphs to illustrate the text.

Summary

In this chapter you have learned:

- Urban land use zones and their characteristics.
- Recent developments in the CBD, inner city, rural/urban fringe in developed world cities with special reference to Glasgow.
- Issues in slum housing in developing world cities.
- Recent developments which deal with issues in slum housing in developing world cities with special reference to Rio de Janeiro.

You should have developed your skills and be able to:

- Locate on a map named examples of developed and developing world cities.
- Identify and locate different land use zones and patterns on an OS map.
- Use research skills to collect and interpret geographical information.

End of chapter questions

National 4 questions

(a) List **four** features that you would find in the **CBD** of a city.

(b) Why does the CBD have the highest land values in a city?

(c) Name the zone built **between** the CBD and the old housing zone.

(d) Why was the old housing zone built right **beside** the old industrial zone?

(e) Where is the new business district located in an urban area **and** why is it found there?

(f) Name an area of slum housing in the developing world.

(g) List **three** issues in an area of slum housing that you have studied in the developing world.

(h) Why does disease spread easily in slum housing?

(i) What is involved in a self-help housing scheme?

(j) How can charities help to improve the standard of living of slum dwellers?

National 5 questions

(a) Describe **and** explain the location of the CBD in an urban area.

(b) Compare the urban environments of an old housing area and a new housing area.

(c) Give reasons for the location of the new industrial zone.

(d) Discuss the shopping changes in the CBD of Glasgow in recent years.

(e) Give examples of **two** transport changes and discuss their impact on an urban area.

(f) Describe the changes that have taken place in the inner city of Glasgow in recent years.

(g) Outline recent developments in the rural/urban fringe area of a city.

(h) Describe **two** different ways of dealing with issues in an area of slum housing that you have studied in the developing world.

(i) Explain why charity work is important in slum housing in developing world cities?

National 5 exam-style questions

You can find sample answers to these exam-style questions on the Leckie website: https://collins.co.uk/pages/scottish-curriculum-free-resources

1 **Study Diagram Q1:** land use zones in an urban area.

Explain in detail why there are different types of housing in the inner city and suburban areas of a city.

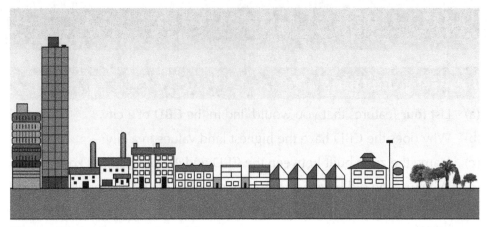

Diagram Q1: *land use zones in an urban area*

(6 marks)

2 What impact does urban sprawl have on rural areas?

(4 marks)

3 **Describe** the measures taken to improve the quality of housing in slum housing.
   You must refer to a **named developing city** that you have studied.

(5 marks)

4 Study the OS map extract of Coventry below.
   Give the 4-figure grid reference of the CBD **and** list the features which show that this area is the CBD.

(4 marks)

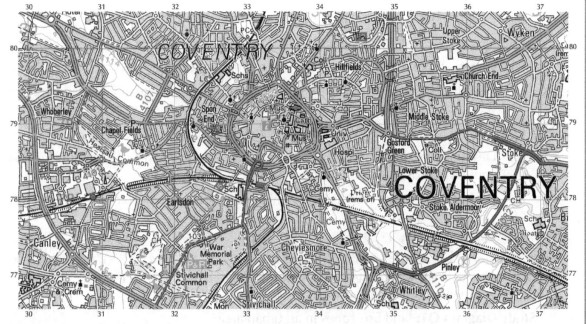

Diagram Q4: *OS map extract of Coventry, scale 1:50 000*

LEARNING CHECKLIST

Now that you have finished the **Urban** chapter, complete a self-evaluation of your knowledge and skills to assess what you have understood. Use traffic lights to help you make up a revision plan to help you improve in the areas you identified as red or amber.

- Locate on a map named examples of developed and developing world cities.

- Describe and explain the location of the CBD in an urban area.

- Describe and explain the location of the old industrial zone in an urban area.

- Describe and explain the location of the old housing zone in an urban area.

- Describe and explain the location of the new housing zone in an urban area.

- Describe and explain the location of the new industrial zone in an urban area.

- Describe and explain the location of the new business district in an urban area.

- List named examples of different land use zones in Glasgow.

- Outline recent developments in the CBD of an urban area in the developed world including:

  ➤ shopping

  ➤ transport

  ➤ spread of CBD into inner city areas

- Discuss recent developments in the inner city of an urban area in the developed world.

- Outline recent developments on the rural/urban fringe of an urban area in the developed world.

- Discuss the issues in slum housing in cities in the developing world including:

  ➢ poor-quality housing

  ➢ overcrowding

  ➢ poor health care

  ➢ unemployment

  ➢ pollution

  ➢ crime

  ➢ landslides

- Describe recent developments which deal with issues in slum housing in the developing world including:

  ➢ self-help housing schemes

  ➢ local authority site and service schemes

  ➢ new towns

  ➢ the work of charities

  ➢ Eviction and demolition programs

- Identify and locate different land use zones and patterns on an OS map.

- Use research skills to collect and interpret geographical information on an urban area.

Glossary

Amenities: Facilities such as shops and services.

Bus lane: A part of the road that is only used by buses.

Business park: An area of modern factories located on the outskirts of a settlement (also called an industrial estate).

Central business district (CBD): The centre of a town or city (also called the city centre).

Charity: A voluntary organisation that helps people without getting anything in return.

Commuters: People who live in one area and travel to another place to work.

Competition: When different companies share the same customers and are rivals for their custom.

Congestion: When there are too many cars or people in one area.

Congestion charges: When people pay money to drive their vehicle into a city centre.

Conservatory: A glass structure built onto the back of a house.

Council estates: An area of public or social housing built by the local council.

Crescent: A street that has a half-moon shape.

Cul-de-sac: A street that is closed at one end (also called a dead-end).

Decentralisation: When businesses move from the city centre to a less congested out-of-town location.

Derelict building: A building that has been abandoned and is empty.

Derelict land: An area of land that has been previously built on and is no longer used.

Detached: A house that stands alone.

Developing country: A country where people have a low standard of living.

Dockland: The area surrounding docks (a place where ships are located).

Double yellow lines: Streets are painted with two yellow stripes to restrict cars parking on them.

Favela: An area of very poor-quality housing in Brazil (also called a slum or slum housing).

Graffiti: Writing or graphics that have been written or sprayed illegally on a surface or wall in a public place.

Greenbelt: An area of land surrounding a town or city which separates settlements and prevents them from joining together.

Greenfield site: An area of land that has never been built on.

Grid iron street pattern: When streets run parallel to one another in different directions (like a grid).

Health centre: A building where doctors and nurses provide health care and advice to the public.

Health care: The resources available to maintain the condition of human beings.

Heavy industry: Businesses that make bulky or heavy goods, e.g. shipyards and steel works.

High density: When there are a lot of things, e.g. buildings, in one place.

Industrial estates: When factories are grouped together.

Industry: The work that people do or business activity.

Informal sector jobs: 'Cash in hand' jobs where people do not pay taxes to the government.

Inner city: The part of a city consisting of the old industrial zone and the old housing zone.

Land use conflicts: A disagreement over the way an area of land is used.

Land use zones: Areas with certain characteristic features within a town or city.

Land value: The cost of a piece of land.

Landscaping: When trees, shrubs and grass are planted in an area to improve its appearance.

Landslide: When the soil and rock on a hillside collapse and travel downhill.

Low density: When houses are detached or semi-detached with space for gardens and garages.

Market: The customers who buy goods and services.

Multi-storey car parks: A place where there are many floors of parking spaces under one roof.

New business district: An area of large shops, supermarkets and entertainments located on the outskirts of a city.

New housing zone: An area of modern housing located on the outskirts of a town/city.

New industrial zone: An area of modern factories located on the outskirts of a town/city.

New town: A planned settlement built to house people from overcrowded inner city areas, e.g. Cumbernauld.

One-way street: Street where vehicles are only permitted to drive in the one direction.

Out-of-town shopping centre: A group of shops all located together under one large roof on the outskirts of a town/city.

Park and ride schemes: People park their car at a bus or train station and travel into town using the bus or train.

Parking charges: When people pay money to park their vehicle.

Pedestrianised streets: Streets that are designated for people to walk and vehicles are not allowed.

Poverty: When a person lacks a certain amount of money or material possessions.

Public transport: Trains, buses and taxis which are available for the public to use for a charge.

Redevelopment: Any new construction on land that has had pre-existing uses.

Refuse collection: The gathering and disposal of household rubbish.

Regenerate: To improve an area, e.g. by upgrading housing.

Renovation: When houses are improved.

Retail park: A group of large shops which sell bulky goods such as electrical goods.

Retailer: Company which sells goods to people.

Ring roads: Roads which bypass city centres to reduce congestion.

Rural land users: People who use the land in the countryside, e.g. farmers.

Rural/urban fringe: The boundary between a town or city and the countryside.

Rural–urban migration: When people move from the countryside to a town or city.

Sanitation: The treatment and proper disposal of sewage.

Scenery: The appearance of a place, especially the natural landscape.

Self-help scheme: A type of informal program that enables people to help themselves and their community to improve houses in a an area of slum housing.

Semi-detached: A house that is attached to another house.

Services: A number of activities that serve the general public for different purposes, e.g. restaurants and cinemas.

Shanty towns: A large settlement consisting of very poor-quality housing.

Shipyards: A factory that builds ships.

Site and service scheme: A program run by the local council in Brazil where new houses are built with running water, toilets and electricity.

Slum: An area of very poor-quality housing (also called a favela in Brazil).

Standard of living: How well off the people of a country are in terms of wealth, health and education.

Subsistence farming: When people only grow enough food to feed themselves and their families.

Suburbs: The outskirts of a town or city.

Tenements: 4/5 storey flats built from sandstone.

Terraced housing: Rows of houses all joined together.

Trading estate: A group of factories located on the outskirts of a town (also called an industrial estate).

Traffic wardens: People who are paid to ensure that drivers do not illegally park their car.

Transport routes: Different forms of transportation including roads, railways and ferry routes.

Tropical storm: An extreme type of weather with very strong winds and heavy rainfall.

Unemployment: When people do not have a job.

Urban regeneration: When an area has been completely transformed by the refurbishment of the buildings and landscape.

Urban sprawl: When towns and cities grow outwards and new developments encroach on the greenbelt.

Vandalism: The intentional and malicious damage to property.

Waste ground: An area of land that has been previously built on and is no longer used.

Zone of transition: An area in a city that is undergoing huge changes (also called the inner city).

9 Rural

Within the context of rural areas, you should know and understand:

- Changes in the rural landscape in developed countries related to modern developments in farming such as: diversification, impact of new technology, organic farming, genetic modification (GM) and current government policy.
- Changes in the rural landscape in developing countries related to modern developments in farming such as: GM, impact of new technology and biofuels.

You also need to develop the following skills:

- Locate on a map named examples of farming areas in developed and developing countries.
- Label the different features of farming landscapes.
- Use research skills to collect and interpret geographical information.

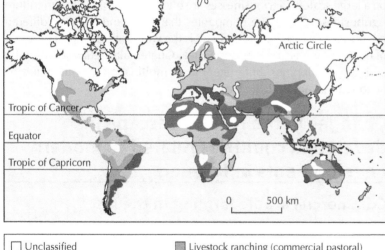

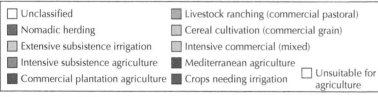

Figure 9.1: *World farming types*

Make the Link

Think about the connection between the weather and different farming types.

CLASSIFYING FARM TYPES

By what we get out of them

Commercial farming produces food for sale

Subsistence farming produces only enough food for the farmer and his family (with perhaps a little left over to sell)

By what we put into them

Intensive farms use large amounts of money, machines and technology or workers

Extensive farms have similar inputs but usually use more land

By what is grown

Arable farms grow crops

Pastoral farms rear animals

Mixed farms grow crops and rear animals

Figure 9.2: *Classification of farming types*

Introduction

The main feature of commercial farming is that it is carried out to make a profit. In richer countries this is often done by individual farmers. On the grassland prairies of the USA and Canada, for example, farmers own huge areas of land and grow wheat or have cattle ranches with thousands of animals. The land use is extensive, because it takes place over a large area, and the farm needs big inputs of money, machinery and fertilisers and pesticides. Workers are often employed only for brief periods of the year when harvesting or planting needs to be carried out quickly.

In poorer countries, most farmers do not have enough money to farm on a large scale so **agribusinesses** are often in control of the agriculture. Agribusinesses are huge companies that own land in many different countries. Del Monte, for instance, is the world's third largest producer of bananas. It has farms in Costa Rica, Guatemala, Ecuador, Colombia and the Philippines but, like many multinational companies, its headquarters are in the USA.

> ## 🔍 HINT
> Think about the type of farming activities that take place in your area.

Changes in the rural landscape in developed countries related to modern developments in farming

Commercial arable farming in the UK

There are many pressures on global food supplies such as climate change, natural disasters and world population growth. Farming is an important **industry** in the UK because it enables us to produce the majority (approximately 70%) of our own food. This is vital so that we do not need to rely on other countries to make and export food to us. If food supplies were to become scarce and exports on the **world market** decreased, the UK would be able to supply enough food to feed its population.

The most important area for **commercial arable farming** in the UK is **the Fens in East Anglia** (SE England). There are over 4000 farms in this area involved in **agriculture**. The Fens is a low-lying area with fertile loam

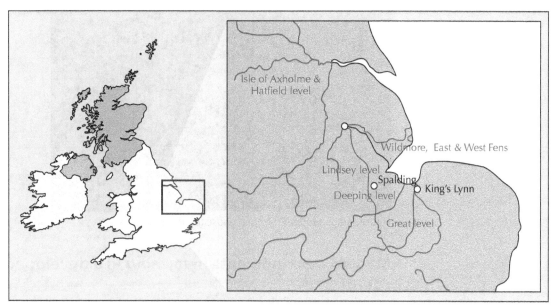

Figure 9.3: *The location of the Fens in the UK*

soils, a warm climate, enough rainfall, good drainage and plenty of sunshine. This makes farming conditions perfect for growing crops like wheat, barley, peas and sugar beet. These crops are produced on a large scale for **profit** by huge farms called **agribusinesses**. Some Fenland farms are considered to be the largest and wealthiest in Europe. They are run like other businesses with managers and workers. Towns such as Spalding and King's Lynn are market towns providing famers with the goods they require such as **machinery**, **seeds** and **fertilisers**. They are also service centres providing farmers with the services they need such as repairs to machinery and accountants.

An arable farming system

Farming is a primary industry and it can be described as a system with inputs, processes and outputs. These are shown in the table below:

Inputs	Processes	Outputs
Climate	Ploughing	Crops, e.g. wheat, barley and vegetables
Flat land	Sowing	Profit
Fertile soils	Irrigating	
Workforce	Spraying	
Machinery	Harvesting	
Capital (money)		
Chemicals		
Seeds		
Transport		
Farm buildings		

Make the Link

Think about where and why arable farming takes place in the course of a river.

Make the Link

Consider the impact of world population growth on food supplies.

Make the Link

Think about the influence of the weather on an arable farm.

Figure 9.4: *A modern farming landscape in the UK*

Modern developments in farming in a developed country: UK

There have been many changes in the UK's rural landscape in recent years due to modern developments in farming practices. These include:

1. Diversification: generating income through non-farming activities.
2. New technology: using machinery, chemicals and irrigation devices to grow crops.
3. Organic farming: growing crops without using chemicals.
4. Genetically modified food (GM): crops derived from scientifically altered seeds.
5. Current government policy: the influence of the EU and UK governments on farming.

Diversification

Many farmers in the Fens work on a large scale and grow crops to sell for a large profit. Other farmers operate on a much smaller scale and experience difficulties trying to make a living. The Wash Fens Rural Development Programme offers farmers a grant to expand into non-agricultural activities. Several farmers have turned to non-farming activities to help them make more money. This is known as **diversification** and includes:

- renting out cottages as holiday homes
- running bed and breakfast facilities
- setting up camp and caravan sites
- opening petting farms
- organising educational tours
- setting up cafes
- selling local produce in farm shops
- offering adventure activities such as quad biking, paintball and pony trekking.

The variety of activities in the countryside has led to an increase in **tourism** in **rural areas**. In some ways the impact has been positive as farmers have increased their income and improved their standard of

HINT

Think about places that you have visited and activities that you have done in the countryside.

living. Money is also invested in rural areas from increased tax revenues. This helps to improve **local services** such as schools. On the other hand, tourism has had an enormous negative impact on the rural landscape. Increased numbers of people erode the land and leave litter, spoiling the **scenery**. **Tourist facilities** such as caravan sites detract from the natural look of the countryside. Farmers' stone walls are damaged by people climbing over them. **Traffic congestion** causes increased noise and air pollution in rural villages and too many parked cars look unsightly.

The impact of new technology

Farm machinery

Since the 1950s there has been a huge increase in the use of **farm machinery** including **tractors**, **ploughs** and **combine harvesters**. **Artificial irrigation** systems such as **sprinklers** help to keep crops watered. This is particularly important in the Fens as seasonal droughts can be a problem.

The use of modern farm equipment like tractors has increased **crop yields** as work is done faster and more efficiently. This means less physical labour for farm workers and an increased profit for the farmer. Fewer workers are needed which means the farmer has less money to pay out in **wages**. Crops are harvested faster so they are fresher. The decrease in **hedgerows** means that time and money is saved in maintenance.

However, the use of machinery has resulted in smaller farms joining together and as a result there are fewer farms in the countryside. Machines are expensive to buy and costly to repair. Tractors are loud and increase noise pollution in the countryside. They also cause increased air pollution as they run on **fossil fuels**. Fewer jobs are available in rural areas and, as people become **unemployed**, they move away to find work. Shops may close and services may decline which leads to further job losses and derelict buildings. **Animal habitats** are lost as hedgerows and trees are removed to increase field sizes to accommodate large machinery. This has caused a loss of **biodiversity** in the countryside, e.g. fewer birds and small animals like hedgehogs. Fields have less protection from the wind as hedgerows act as natural **shelter belts**. Soil erosion increases as the roots are no longer there to help bind the soil together and hold it in place. Removing trees and hedgerows also has a negative impact on natural scenery as this creates a flat, featureless landscape.

Figure 9.5: *Modern farm machinery*

Chemicals

There has also been an increase in the use of **fertilisers and pesticides** to help improve the quality of crops and increase farmers' profits. Chemical fertilisers also allow crops to be grown on areas with poorer soil quality. However, farm chemicals cause water **pollution** as the run-off is washed into rivers. This can cause algae which use up the oxygen in waterways and affect the wildlife living there.

Organic farming

Of the 77% of the UK's land surface that is farmed, 4% is used for producing food organically. **Organic farming** involves the production of crops without using **artificial chemicals** like fertilisers and pesticides.

This means that water pollution is reduced and aquatic wildlife is protected from harmful run-off from farm chemicals in certain areas. Organic farming relies on more natural farming techniques such as **crop rotation** and using animal **manure** as a fertiliser. Many farmers are turning to this type of farming as **consumers** choose to buy **chemical free food**. Organic farming is more intensive as it takes more time and effort to grow crops. Productivity is lower as crops cannot be mass produced on a large scale and the soils have not been chemically improved. As a result, the cost of **organic produce** is higher to ensure a decent income for organic farmers.

Genetic modification (GM)

Genetic modification **(GM)**, such as rice and wheat, is grown from seeds which have been altered by modern science. Traditionally, plants were improved by farmers cross-breeding them with better plants. This allowed farmers to grow sturdy plants which produced large amounts of crops. This natural process takes years to improve crop varieties but GM plants are altered artificially in a lab by scientists at a much faster rate. Altering plants' genetic make-up seems to go against nature and many people are opposed to GM crops. Some countries have banned the growth of GM crops until more research has been conducted into their impact on people's health and the environment. On the other hand, scientists believe that GM technology is the future of food production. Some experts argue GM will prevent the world's population from starving because plants can be bred to survive pests, diseases, weeds, drought and frost and so will prevent crops from being wasted.

Current government policy: EU and UK

EU

The European Union's **Common Agricultural Policy (CAP)** was set up to improve the standard of living of farmers in Europe and help make individual countries' food supplies more secure. The main aim of the CAP is to improve agricultural productivity so that consumers have a stable and affordable food supply.

- The EU provides farmers with income support to reimburse them for maintaining the landscape, soil and wildlife in the countryside.
- Farmers are also given financial assistance to work in an **environmentally friendly** and **sustainable** way.
- **Grants** are provided to help farms modernise their business, e.g. money is provided to buy new machinery and upgrade farm buildings.
- **Loans** are given at low interest rates to assist various diversification (non-farming) projects.
- Advice and training programmes are provided for farmers to keep them up-to-date with the latest policy and practice in agriculture.
- **Subsidies** are given to improve farmers' income as they have to compete with foreign imports.

UK

As well as the EU, the UK government can also influence the rural landscape. The government, farmers and land managers are working together to protect and manage important habitats and to improve the landscape and wildlife in the countryside.

- Subsidies are given for Quality Assurance Schemes which reward farmers for producing top quality food, e.g. prime beef.
- Farmers are paid to ensure animal welfare and food safety.
- Woodland grants are given to farmers to plant and maintain trees.
- Grants are provided for activities which help to sustain rural communities, e.g. farmers creating jobs for local people.
- **Countryside Stewardship Schemes** offer farmers extra income for protecting and maintaining rural areas, e.g. planting hedgerows.
- **Rural Development Contracts** pay farmers for various activities such as improving water quality and tackling climate change.

The Great Fen Project is financed by many organisations including the Environment Agency, local councils and lottery funding. It involves the restoration of the traditional Fenland landscape by flooding small areas of land to create nature reserves and open-water habitats to encourage wildlife. A mosaic of natural habitats including woodland, bogs, dry meadows and natural fens have been created from arable fields. This has been done to increase wildlife and improve the natural look of the area.

Make the Link

Think about how rural areas in coastal and limestone landscapes are protected.

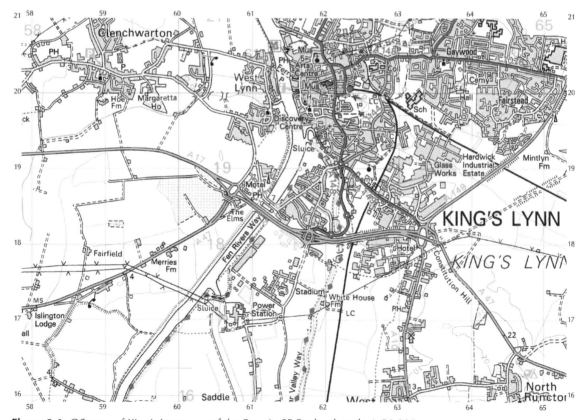

Figure 9.6: *OS map of King's Lynn area of the Fens in SE England, scale 1:50 000*

GO! ## Activity 1: Group activity

1. In groups, investigate how government policy affects other types of farming in the UK. You should research current policies which affect dairy, hill sheep and beef cattle farming.
2. Write a podcast detailing your findings using appropriate language.
3. Decide who will be the presenter.
4. Film the person making the speech.
5. Upload it onto your personal space on the internet.

Make the Link

In the Population chapter, you learned about indicators of development: the number of calories consumed per person per day indicates the standard of living in a country.

Changes in the rural landscape in developing countries related to modern developments in farming

Introduction

Agriculture is an extremely important industry in developing countries as many nations cannot afford to pay to import the food they need to feed their population. Worldwide, 37% of people live in villages and work on farms. The majority of these farms are located in developing countries because their economies have not advanced and people have to grow their own food for survival.

Intensive subsistence farming in India

Figure 9.7: *The location of the Ganges Valley in India*

Farming is the largest industry in India with more than 52% of the total population working on farms and living in villages. This type of farming is known as Intensive Subsistence Agriculture because it involves people growing food for themselves. With over 1·24 billion people, India has the second highest population in the world to feed. Rice is the staple food of 65% of the population of India and makes up 90% of the total diet.

Fruit trees to add to diet

Small fields (called paddies) which may get smaller when passed on to the next generation

Animals may be used for ploughing instead of expensive machines

Small rice plants may be grown in nursery beds before being moved to bigger fields

Low walls (called bunds) built to keep water in the paddy fields

Fish may be added to paddy fields to add protein to diet

Rice planted under water from heavy rainfall, flooding or irrigation

Figure 9.8: *Features of the rice farming landscape*

An important area for rice farming is the Lower Ganges Valley in NE India. This area is one of the most densely populated regions in the world. The warm, wet climate and large area of flat land is ideal for wet rice cultivation. Traditional farming involves small farms and **manual labour**. Field sizes are small, usually 1 hectare (the size of a football pitch) and sub-divided into about 15 plots. Due to inheritance laws, land is divided among a farmers' sons, so fields become smaller and smaller. Fields are surrounded by mud embankments called bunds which are built to hold water in the flooded fields. Fish are often added to the paddy fields to provide a source of protein for farmers. The annual flooding of the River Ganges provides silt which helps to fertilise the soil. There are few farm machines as farmers cannot afford them, so they use **water buffalo** to plough the fields and manual labour to **harvest** the crop. As rice requires a constant supply of water, fields are often constructed beside a water supply such as the River Ganges or man-made irrigation canals. Hillsides are **terraced** to make even more flat land to grow crops as farmland is so important.

Make the Link

Figure 9.8 represents the geographical skill: annotating a photograph.

Figure 9.9: *A terraced hillside with rice fields in India*

Modern developments in farming in a developing country: India

There have been many changes in India's rural landscape in recent years due to modern developments in farming practices, which include:

1. The impact of new technology
2. Genetically modified food (GM)
3. Biofuels.

Impact of new technology

India is a developing country with a rapidly growing population to feed. It cannot afford to import all the food it needs so many people have to grow their own. Since the 1960s, the introduction of new farming practices called the **Green Revolution** have helped to secure food supplies for the growing population. It has included the introduction of machinery, irrigation channels, fertilisers, pesticides, improved varieties of seeds (see section 2 on GM) and better infrastructure. The government also introduced a Land Reform Policy.

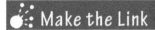

Make the Link

Consider the impact of rural–urban migration on both rural and urban areas.

- **Machinery:** tractors are used to plough fields and harvest crops. Trucks and trailers have replaced bullock carts to transport crops. However, tractors increase noise and air pollution. Rural unemployment has also increased so people are moving to cities.

- **Irrigation channels:** ditches have been dug to store water for use in the dry season. This allows two to three harvests per year due to a continuous water supply. However, this means that soil is exhausted as it is in constant use.

- **Fertilisers and pesticides**: fertilisers are spread onto fields to improve crop growth and pesticides are sprayed onto crops to protect them from diseases. However, they pollute nearby rivers and contaminate local drinking water.

- **Land reform**: field sizes were made bigger by joining smaller plots together to accommodate machinery and make it easier for farmers to manage their land.

- **Improved infrastructure:** roads and railways have been improved to export crops to market quickly. New roads allow farm machinery to be brought into the region more easily and manoeuvred around it.

The Green Revolution

Aims: to scientifically develop new types of rice that give higher yields; to introduce new irrigation schemes; to use new fertilisers to help crops grow and new pesticides to protect them from disease.

Benefits	Problems
• Food production has increased, so more people can be fed.	• Not all farmers have felt the benefits of the Green Revolution.
• Higher yields mean lower prices, so poor people can afford more food.	• The less well-off who could not compete have sold their land and moved to the city.
• Crops are more resistant to disease, so harvests are more reliable.	• Some farmers have borrowed money to pay for crops and are now in debt.
• Crops grow faster, so more harvests can be gathered each year.	• Machinery has caused rural unemployment.
	• Chemicals have polluted local water supplies.
• The better-off farmers who can afford chemicals and machinery have become richer and employ more local people.	• Irrigation has increased the demand on drinking water stores.
• More jobs are available in businesses supporting farming.	• Some of the new varieties of rice are not as pleasant to eat.

Figure 9.10: *Benefits and problems of the Green Revolution*

Impact of intermediate technology

Many people in poor developing countries like India cannot afford expensive equipment for growing food and rely on more **appropriate (intermediate) technology**. This involves the use of simple and inexpensive solutions that use the existing skills of local people and include:

- Building **dams** on seasonal rivers to collect extra water.
- Making **terraces** on sloping land to limit soil erosion.
- Using **solar energy** to pump water from wells for irrigation.
- Constructing **stone lines** across fields to trap soil and water.

Genetic modification (GM)

As part of the Green Revolution, **High Yield Variety (HYV)** rice seeds were developed by scientists to improve food supplies and eliminate **famine** in developing countries. Some HYV or 'miracle' seeds can produce up to ten times more crops than regular seeds on the same area of land. They can also be grown in areas that were previously unsuitable for growing crops. HYV crops are shorter in height so more able to withstand high winds and heavy rain. More food is grown on an area of land which increases farmers' profits. However, HYV crops need a lot of fertilisers and pesticides to grow, which increases costs and pollution. They also require a more reliable source of water and irrigation increases farmers' costs.

> ### ᯤ: Make the Link
> Think about the purpose of growing GM crops in developed and developing countries.

Biofuels

Biofuels are a source of fuel made from plants or waste. They are environmentally friendly, sustainable and an alternative to fossil fuels (coal, oil and natural gas) which are more expensive than biofuels and pollute the environment. Despite this, many industries are reluctant to seriously invest in the development of biofuels while fossil fuels are still available. In India, Jatropha plant seeds which contain more than 40% oil are **cultivated** and processed as a biofuel. The oil is used after extraction because it does not need to be refined. The oil (biodiesel) is used to power diesel generators and engines in machinery, e.g. tractors. Biofuels have many benefits such as **employment** of people to grow and process the plants. Trees are protected as they do not need to be cut down to provide fuel. This helps to reduce soil erosion as trees bind the soil together and keep it in place, allowing crops to be grown. The plants used to make biofuels grow in dry marginal areas which are not used for farming, so no valuable farmland is lost in India.

> ### ᯤ: Make the Link
> Think about the impact of natural disasters on farming in developing countries.

🔘 Activity 2: Paired activity

1. In pairs, read the section titled Impact of Intermediate Technology.
2. Write a homework question for each other – one question should begin with the word **describe** and the other, **explain**.
3. Note down the question that you have to answer in your homework jotter.
4. Answer this question at home before the next lesson.
5. Be prepared to share and discuss your answer with your partner and group.

GO! Activity 3 (National 5)

Individually, research and write an illustrated essay comparing the differences in farming in the UK and India. You should include:

- A world map showing the location of the UK and India.
- A description of the different types of farming that takes place in both countries.
- A labelled sketch of the farming landscapes.
- An outline of any **two** modern developments for both countries.
- An explanation of the impact of modern developments on the rural landscape.
- **Two** graphs showing data of your choice, e.g. two pie charts comparing the income generated from farming in the UK and India.

Summary

In this chapter you have learned:

- Types of farming and their location.
- Classification of farming types.
- Changes in the rural landscape in developed countries related to modern farming developments: diversification, impact of new technology, organic farming, Genetic modification (GM) and current government policy.
- Changes in the rural landscape in developing countries related to modern farming developments: GM, impact of new technology and biofuels.

You should have developed your skills and be able to:

- Locate on a map named examples of farming areas in developed and developing countries.
- Label the different features of farming landscapes.
- Use research skills to collect and interpret geographical information.

End of chapter questions

National 4 questions

(a) Name an important area for commercial arable farming in the UK.

(b) Give **two** reasons why farming is important in the UK.

(c) List **four** things that make the Fens the most important agricultural area in the UK.

(d) Look at Figure 9.4: a modern farming landscape in the UK. Draw a sketch and label the features of this landscape.

(e) Outline **three** modern developments in farming in developed countries like the UK.

(f) What is diversification **and** why is it important for farmers in developed countries?

(g) Name an area of wet rice cultivation in SE Asia.

(h) List **six** things that are required for wet rice cultivation.

(i) Describe **three** features of the Green Revolution.

(j) Explain the importance of genetically modified food.

National 5 questions

(a) **Describe, in detail,** the location of different farming systems throughout the world.

(b) **Describe** at least **two** modern farming developments in the UK.

Choose from: diversification, new technology, organic farming, GM and current government policies.

(c) **Explain** the impact of modern developments in farming in **developed** countries on **both** people **and** the farming landscape.

(d) **Give reasons** for the importance of farming in developed and developing countries.

(e) **Describe two** ways that both EU **and** UK government policy affects farmers.

(f) **Describe** the location of Merries Farm (Grid Reference 5917) in Figure 9.6.

(g) **Explain** the impact of modern developments in farming in developing countries on both people **and** the landscape.

(h) **Explain** the importance of the Green Revolution.

(i) Outline the importance of intermediate technology in developing countries.

(j) **Describe** the benefits of biofuels.

National 5 Exam-Style Questions

You can find sample answers to these exam-style questions on the Leckie website:
https://collins.co.uk/pages/scottish-curriculum-free-resources

1 **Diagram Q1: Recent changes in agriculture in developing countries**

INCREASED USE OF FERTILISERS	USE OF TRACTORS	IRRIGATION CHANNELS

Describe the benefits **and** problems that changes such as those shown in the diagram above have brought to developing countries.

(5 marks)

2 **Diagram Q2: Recent changes in British farming**

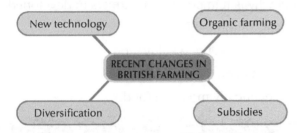

Choose **two** of the recent changes in farming.

Explain, in detail, why these bring both benefits **and** problems to the farmer.

(5 marks)

3 **Explain** how farmers in developing countries such as India have been able to increase crop outputs.

(4 marks)

4 **Explain** the changes in the rural landscape in developed countries, related to modern developments in farming.

Your answer **must** refer to a country that you have studied.

(4 marks)

LEARNING CHECKLIST

Now that you have finished the **Rural** chapter, complete a self-evaluation of your knowledge and skills to assess what you have understood. Use traffic lights to help you make up a revision plan to help you improve in the areas you identified as red or amber.

- Locate on a map named examples of farming areas in both developed and developing countries.

- Outline the classification of farming types.

- Explain the importance of farming in developed countries like the UK.

- List six things that make the Fens the most important agricultural area in the UK.

- Describe the impact of diversification on the farming landscape in a developed country.

- Describe the impact of modern technology on the farming landscape in a developed country.

- Describe the impact of organic farming on the farming landscape in a developed country.

- Describe the impact of genetic modification on the farming landscape in a developed country.

- Describe the impact of current government policy on the farming landscape in a developed country.

- Outline the importance of farming in developing countries like India.

- Describe the impact of new technology on the farming landscape in a developing country.

- Describe the impact of genetically modified seeds on the farming landscape in a developing country.

- Describe the impact of biofuels on the farming landscape in a developing country.

Glossary

Agribusiness: A large-scale farming business that operates to make a big profit.

Agriculture: The cultivation of crops or animals for food (also called farming).

Animal habitats: The areas where animals live.

Appropriate (intermediate) technology: Simple measures undertaken to improve farming in poor countries.

Artificial chemicals: Man-made substances that are used in farming to grow healthy crops.

Artificial irrigation: When crops are artificially watered by sprinklers and irrigation canals due to a lack of rainfall.

Biodiversity: The variety of plants and animals in an area.

Biofuels: Any kind of fuel made from living things, e.g. plants or waste.

Chemical free food: Food that has been produced without the use of fertilisers and pesticides.

Combine harvester: A farm machine used to gather cereal crops such as wheat from fields.

Commercial arable farming: Growing crops on a large scale for sale.

Common Agricultural Policy: A set of European Union laws that impact on farmers.

Consumers: People who buy goods.

Countryside: A rural area.

Countryside Stewardship Schemes: A scheme passed by the government that pays farmers to protect and maintain their land.

Crop rotation: When fields are used to grow the same crop in alternate years. This helps to maintain soil fertility.

Crop yield: The amount of crops harvested from a certain area.

Cultivated: When land is used for growing crops.

Dam: A large concrete wall built to hold water in a reservoir.

Developed country: A country where people have a high standard of living.

Diversification: When farmers undertake non-farming activities to make more money.

Educational tours: When people are guided around and given a talk about the working life of a farm.

Environmentally friendly: Activities which are not harmful to the environment.

Employment: When people have a job.

Enterprise: Undertaking activities that make money.

European Union (EU): An organisation consisting of 28 countries joined by an economic and political agreement.

Famine: A widespread lack of food caused by different factors including drought (lack of rainfall).

Fertile soils: Soils which are rich in nutrients.

Fertilisers: Chemicals which help crops to grow.

Fossil fuels: Fuels which are finite, e.g. oil, coal and natural gas.

Genetically modified food (GM): Food that has been grown from seeds which have been altered by science.

Government policy: The government's plan of action and the regulations and laws passed by them that must be followed.

Grant: Money that is given and does not have to be paid back.

Green Revolution: The application of modern farming techniques, e.g. fertilisers, HYV seeds and irrigation.

Habitats: The areas where plants and animals live.

Harvest : To gather in crops from the fields.

Hedgerows: Rows of small trees and shrubs shaped into a wall of vegetation.

High Yield Variety (HYV) seeds: Seeds which can produce up to ten times more crops than regular seeds.

Irrigation: When crops are artificially watered by sprinklers and irrigation canals.

Industry: The type of work that people do or business activity.

Infrastructure: The framework for allowing people to exist in a place, e.g. roads, houses and electricity.

Irrigation channel: A man-made ditch used to collect rainwater.

Land reform: When land is redistributed to landless people.

Loan: Money that is given and has to be paid back with interest.

Local services: A number of activities that serve the general public for different purposes, e.g. schools and shops.

Machinery: Mechanically operated equipment, e.g. tractor.

Manual labour: When people do all the work by hand.

Manure: Animal dung used as a natural fertiliser.

Market: The customers who buy goods.

Mechanisation: The use of farm machinery.

Migration: When people move from one place to another.

Organic farm: A farm which produces food without using chemicals.

Organic farming: When farmers produce food without using chemicals.

Organic produce: Food that has been produced without the use of chemicals.

Pesticides: Chemicals which kill pests that attack and eat crops.

Plough: A farm machine used to turn over the soil in preparation for seeds being sown.

Pollution: When the air, water or land is damaged by harmful chemicals.

Poverty: When a person lacks a certain amount of money or material possessions.

Profit: The amount of money a farmer makes after all expenses have been paid.

Recreation: An activity undertaken during leisure time.

Rural area: An area of countryside.

Rural depopulation: When people move away from the countryside.

Rural Development Contracts: An agreement between farmers and the government which ensures payment to farmers for managing and improving the environment.

Rural landscapes: Areas of countryside and farmland.

Scenery: The appearance of a place, especially the natural landscape.

Services: A number of activities that serve the general public for different purposes, e.g. schools and shops.

Shelter belts: Rows of trees or hedges planted to protect crops from being damaged by the wind.

Solar energy: Energy generated from the heat of the sun.

Sprinkler: A spray used to water crops artificially.

Starvation: When people are extremely hungry due to lack of food over a long period of time.

Stone lines: Rows of stones laid across a field to trap soil and water so that crops can be grown.

Subsidies: Money that is paid to farmers to top up their income.

Subsistence farming: When farmers only grow enough food to feed themselves and their families.

Surplus: When there are extra crops.

Sustainable: When a resource can be used over and over without it being totally used up or destroyed.

Terraced hillsides: When a hill has steps cut into the side to make more flat land to grow crops.

Tourism: Travel for recreation or business purposes.

Tourist facilities: Services provided for visitors, e.g. toilets.

Tractor: A machine used to do various types of work on a farm.

Traffic congestion: When a large volume of vehicles cause traffic jams.

Unemployed: Be out of work and not have a job.

Wages: The amount of money that people earn.

Water buffalo: A large animal used to pull simple farm ploughs in poor countries where farmers cannot afford tractors.

Wetlands: An area where the land is marshy.

Woodland: An area of trees (also called a forest).

World market: Countries across the globe which are able to buy resources.

Yield: The amount of crops harvested from a certain area.

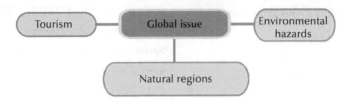

❖ The study of tourism enables us to know and understand the impact of visitors on different areas locally, nationally and internationally. By analysing how tourism affects different landscapes and local people, it allows us to consider ways to manage the impact of tourism.

❖ By looking at how earthquakes, volcanoes and tropical storms impact on the landscape and people, it enables us to consider different ways to manage their impact through prediction and planning.

❖ Sustainable development is a key theme in Geography. It is essential to study the impact of human activity on the natural environment as it allows us to understand the effects of the degradation of both rainforest and tundra environments. By looking at how we impact on these landscapes, it allows us to consider different ways to protect them for future generations.

Consider how you currently help to protect the environment.

Think about environmental hazards that you have seen on the news.

Reflect on holiday destinations and the impact that visitors have on the environment.

Global Issues

10 Natural regions

> **Within the context of rainforest and tundra environments, you should know and understand:**
>
> - A description of the climates and ecosystems.
> - Their use and misuse by people.
> - The effects of degradation on people and the environment.
> - Management strategies to minimise the impact/effects.

> **You also need to develop the following skills:**
>
> - Locate on a map the location of rainforest and tundra environments.
> - Draw and interpret climate graphs.
> - Extract, interpret and present numerical and geographical information which may be statistical, graphical or tabular.

The equatorial climate region

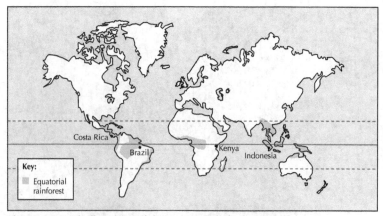

Figure 10.1: *The location of the equatorial climate region*

Equatorial or **tropical rainforests** are found near the **Equator**. Places which experience an **equatorial climate** include parts of Central America, e.g. Costa Rica; South America, e.g. Brazil; Central Africa, e.g. Kenya; and South-East Asia, e.g. Indonesia.

Figure 10.2: *An equatorial rainforest landscape*

The equatorial climate

The equatorial climate region is hot and wet all year round. The **annual rainfall** is high as it usually rains every day. Temperatures are almost the same all year, around 28°C. The **temperature range** is usually only a few degrees. There are no seasons and **humidity** is high in the equatorial climate region.

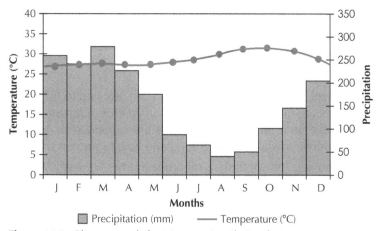

Figure 10.3: *Climate graph for Manaus, Brazil (South America)*

> ### 🔍 HINT
>
> The bars indicate rainfall in every month and the line is almost straight, indicating that temperatures are much the same throughout the year.

The equatorial ecosystem

Equatorial rainforests contain a vast number of plants and animals, about 90% of the world's **species**. **Vegetation** is dense and occurs in four main layers from top to bottom:

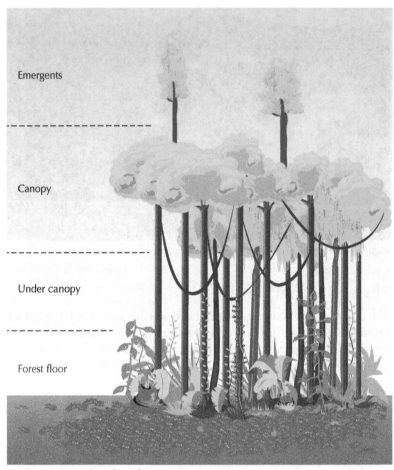

Emergents

Canopy

Under canopy

Forest floor

Figure 10.4: *Layers of rainforest vegetation*

1. The **emergents** are the tallest trees and can grow up to 60 metres tall. They are higher because they absorbed more sunlight to help then make more food. Emergent trees are supported by **buttress roots** which grow above the ground and help to support them.

2. The **canopy** is the continuous umbrella of trees between 20 to 40 metres tall. This layer contains fruit and leaves, and is home to most **wildlife** including insects, birds and some mammals such as monkeys, jaguars and sloths.

3. The **under canopy** or **under storey** is shady and cooler as there is limited sunlight. Plant **seedlings** lie dormant until larger trees die, and the gap that is left is quickly filled as they grow into it. **Vines** take root in the ground and grow up trees to reach the sunlight.

 HINT

You can use a mnemonic to help you remember the rainforest layers such as:
Everyone (**E**mergents)
Can (**C**anopy)
Use (**U**nder canopy) a
Fork (**F**orest floor)

4. The **forest floor** is dark as the canopy blocks out most of the sunlight. This layer of rainforest is covered in a thick blanket of dead and **decaying** roots and leaves, called **humus**. A little vegetation can grow between the trees if it is able to trap sunlight. This area is likely to flood during heavy rainfall.

The nutrient cycle in the equatorial rainforest

Dead leaves and plant material are quickly broken down by the warm, wet climate and the large numbers of insects on the forest floor. The substance that is made, called humus, provides **nutrients** that are easily absorbed by plant roots. However, nutrients are used up rapidly by the number of plants competing for survival. The **nutrient cycle** is unbroken as long as **deforestation** does not occur. However, if the protective canopy of trees is removed, the soil quickly becomes infertile. An absence of leaves means that humus cannot be made to replace the nutrients in the soil. Heavy rainfall is also a problem as it quickly **leaches** nutrients out of the soil making it infertile.

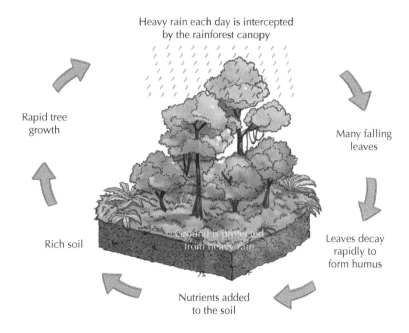

Figure 10.5: *Rainforest nutrient cycle*

The rainforest water cycle

The roots of trees and plants absorb water from the soil. As temperatures increase during the day, water evaporates from the soil and plants' leaves into the atmosphere. This water vapour cools, condenses and forms clouds. The water droplets become bigger and bigger and the clouds eventually burst, bringing the next day's rain. This is known as **convectional rainfall**. The whole process is repeated daily and the cycle continues.

Make the Link

Consider how convectional rainfall is different from the relief and frontal rainfall that were covered in the Weather chapter.

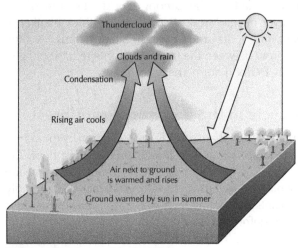

Figure 10.6: *Convectional rainfall*

Uses of the equatorial environment by people

Shifting cultivation

Native tribes have lived in the world's rainforests for thousands of years without destroying it. The **Kayapo Amerindian Tribe** who live in the Amazon rainforest practise a simple form of agriculture called **shifting cultivation**. This is a traditional method of farming and involves growing food for survival. Along with other features of their way of life, shifting cultivation is under threat from large-scale clearance of the rainforests.

A small area of land is cleared using stone axes and **machetes**. The vegetation is burned and the ash is spread over the ground as a natural **fertiliser**. This type of farming is called **'slash and burn'**. Larger trees and those which have fruit are left to help **regeneration**. Crops such as **manioc** and **yams** are planted using a long pointed digging stick in clearings called **chagras**. The diet is supplemented by hunting, fishing and gathering fruit. After two to three years crops no longer grow well as the soil's fertility is exhausted. Nutrients have been used up by the crops and rain washes the goodness out of the top soil by leaching. The tribe has to move on and clear another area of forest to begin the process again. The original area is rejuvenated over 30–60 years, as it receives nutrients and seeds from surrounding vegetation. This use of the rainforest is **sustainable** as vegetation will grow back in time.

Medical research

Many rainforest plants have valuable medicinal properties. More than 60% of **medicines** sold in chemists' shops contain ingredients derived from rainforest vegetation. Plants are removed for scientific and medical research which may provide new food sources as well as medicines for life threatening **diseases**, many of which still remain undiscovered.

Misuse of the equatorial environment by people

The world's equatorial rainforests are being destroyed for many reasons: land is cleared for **new roads** such as the **Trans-Amazonian Highway** in Brazil to improve access. Resources such as **iron ore**, gold, silver and diamonds are exploited to reduce developing countries' **debts** as large **profits** can be made from **mining**. Jobs are created in **industries** such as **logging, cattle ranching, plantations** and **hydro-electric power schemes**. Income is generated when **timber**, **minerals** and other resources are sold on the **world market**. The profits are often used to help many countries improve **infrastructure**, e.g. roads and services such as schools and hospitals.

Effects of degradation on people

Developments such as mining, logging and cattle ranching have resulted in a loss of land for **native people**. Contact with companies from developed countries has led to a loss of their **traditional way of life**. Tribes have been influenced by new technologies such as rifles and radios. Some tribes have been forced into **reservations** for their own protection against new developments by foreign companies. Native people have contracted 'western' diseases such as flu and measles as they have no **immunity** to them. The traditional farming system is being altered as there is less land to move to in order to practise shifting cultivation. As a result, many people are being forced to practise sedentary farming.

Effects of degradation on the environment

Clearing the land for new developments causes deforestation. **Tropical hardwood trees** such as mahogany and teak take hundreds of years to grow so they are difficult to replace. Removing the rainforest threatens the survival of many rare plant, bird and animal species. This could lead to their **extinction** as their **habitat** is destroyed. Fertile but fragile soils are swiftly washed away when trees are cleared. **Flooding** can occur if soil is washed into rivers.

Mining activities cause water **pollution** and negatively affect aquatic life. Vital ingredients for medicines are lost as vegetation is destroyed. Fewer trees to remove carbon dioxide from the atmosphere leads to increased **global warming** and **climate change**. Rainforest areas turn to **desert** due to a breakdown of the water and nutrient cycles.

Figure 10.7: *Logging in the Amazon rainforest*

> ### 🔎 HINT
> Think about how long it takes for a tropical hardwood tree to grow in comparison to the time it takes for a machine to chop it down.

> ### 🔎 HINT
> Think about why developing countries continue to destroy their rainforests despite the fact they potentially hold the key to cures for killer diseases like Cancer.

Figure 10.8: *Illegal gold mining in the Amazon rainforest*

Management – strategies to minimise the impact/effects

Many developing countries such as Borneo, Brazil and Kenya need to exploit their rainforest resources in order to generate income to help them become more developed. Many of them now realise that **exploitation** must be done in a manageable way. To ensure the rainforest is conserved for **future generations**, **sustainable development** is vital for its survival. Management strategies designed to do this include:

1. **National parks**. Areas of rainforest can be designated as a national park by governments passing laws to protect them from **commercial developments**.

2. **Agro-forestry**. This involves growing trees and crops at the same time to allow crops the benefit of being sheltered by trees. It also prevents soil **erosion** as trees' roots bind the soil. Crops benefit from the nutrients provided by the humus from leaves.

3. **Sustainable forestry schemes**. Experts calculate how many trees can be cut down without causing irreparable damage to an area. Trees can be cut down as long as the same number of trees is re-planted.

4. **Selective logging**. Trees are only cut down when they reach a specific height. This helps protect young trees and aids regeneration where deforestation has occurred.

5. **Afforestation**. This involves planting more trees to sustain the canopy of trees.

6. **Monitoring**. **Satellites** are used to take **aerial photographs** of areas of rainforest to check that activities taking place are legal.

7. **Education**. This involves training people who are involved in the exploitation of the rainforest to understand the consequences of their actions.

8. **Eco-tourism**. This can help to protect areas of rainforest as tourists' money is invested in conservation programs.

Make the Link

In the Tourism chapter you learn the benefits and problems of eco-tourism in more detail.

GO! Activity 1: Group activity

1. In groups, investigate:
    (a) Why the rainforest is rapidly disappearing in Borneo.
    (b) The effects of degradation on people and the environment.
    (c) The management strategies aimed at conserving the areas of pristine rainforest that remain.
2. You must include a location map, graphs and photos to illustrate your findings.
3. You should display the information your group has gathered in a PowerPoint presentation.
4. Decide who will talk about each slide and practise your presentation.
5. Present your group PowerPoint presentation to your class.

The tundra climate region

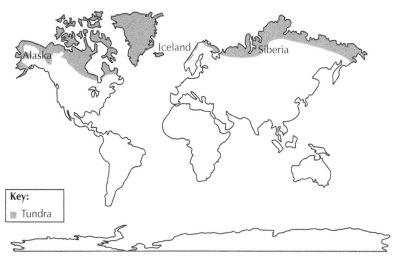

Figure 10.9: *The location of the tundra climate region*

Tundra environments are found in the **northern hemisphere** above the **Arctic Circle**. Places which experience an arctic tundra climate include parts of North America, e.g. Alaska; Europe, e.g. Iceland; and Asia, e.g. Siberia.

Figure 10.10: *A tundra landscape*

The tundra climate

The tundra climate is very cold and dry all year round. Tundra climates are sometimes called **cold deserts** because the annual **precipitation** is less than 250 mm. Temperatures remain below 0°C most of the year. There is a large temperature range as temperatures can rise to 12°C and fall below −70°C.

HINT

Precipitation is any type of moisture falling from the sky and includes rain, hail, sleet and snow.

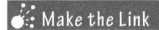

Make the Link

Consider the reasons for the differences between the rainforest and tundra climate graphs, e.g. latitude.

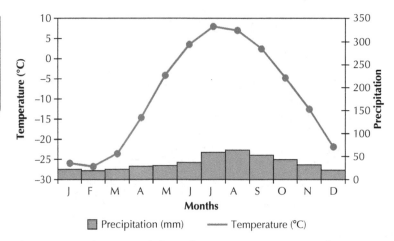

Figure 10.11: *Climate graph for Iqaluit, Nunavut (Arctic Canada)*

The tundra ecosystem

Tundra environments are known as a treeless plain. The very cold and dry conditions mean that these areas contain a small variety of plants and animals, in comparison the rainforest. For most of the year the soil is permanently frozen, known as **permafrost**. During summer, the surface soil (about 50 cm) thaws but underneath remains frozen. Consequently, plants that require deep root systems cannot grow. The harsh climate means that little vegetation grows in the tundra even during summer months when temperatures rise above freezing point. The **growing season** is no more than two months long and the soil is thin and relatively infertile. The bare, rocky ground can only support low-growing plants such as **shrubs**, **mosses** and **lichen**.

Figure 10.12: *Tundra landscape in winter*

During winter the tundra is covered in a layer of snow and ice. Frozen soil, icy winds, severe temperatures and little precipitation mean that vegetation cannot survive. Animals such as the polar bear, arctic fox and snowshoe hare turn white to **camouflage** themselves to help them hunt for food.

Figure 10.13: *Tundra landscape in summer*

During summer, when the snow and top layer of permafrost melt, the landscape is covered with **marshes**, **lakes** and bogs. Water lies on the surface as **evaporation** is slow and the permafrost prevents it from draining away. Animals such as **caribou** (reindeer), birds, e.g. **sandpipers**, and swarms of insects migrate into the tundra to take advantage of seasonal food sources.

Make the Link

Compare evaporation rates and cloud amount in the equatorial and tundra climate regions.

Use of the tundra environment by people

Native people known as **Inuit hunters** from Nunavut in Arctic Canada have lived in the tundra for thousands of years without destroying it. They practise an ancient way of living by hunting animals such as seals and fishing for all their food. It is a sustainable existence which involves catching enough food for survival. Along with other features of their culture Inuit hunters' methods of obtaining food are under threat from large-scale modern developments.

Misuse of the tundra environment by people

The tundra environment is one of the least disturbed **ecosystems** in the world. However, that is changing for many reasons with the discovery of large reserves of raw materials. **Illegal hunting and fishing** are threatening the numbers of certain species such as whales. This is also reducing the food supplies for local people. **Oil rigs** have enabled companies to drill for oil in the sea. In Alaska, oil is transferred from **oil fields** in Prudhoe Bay to the ice-free port of Valdez where it is **exported** by ships. **Mining** is a growing industry and mines have opened up resources under the land in tundra regions, e.g. Arctic Canada. **Natural gas** (methane hydrate) is extracted from the Messoyakha gas field in western Siberia. Natural gas is pumped from beneath the permafrost and piped east across the tundra to the Norilsk metal smelter – the biggest industrial enterprise in the Arctic. New **industries** have led to the creation of **towns** such as Anchorage in Alaska which have been built to provide accommodation for workers. **New roads** have been built to transport people and goods.

Figure 10.14: *An Inuit hunter*

Figure 10.15: *An Alaskan oil pipeline*

HINT

Consider why the pipeline is raised above the ground and not placed underground.

Effects of degradation on people

Illegal hunting and fishing threaten the traditional way of life of hunters as their food sources are endangered. Local people have found jobs in new industries such as mining. This gives them money to improve their **standard of living** but alters their lifestyle. **Entertainments** and modern transport influence the traditional routine of native people. Chemical **pesticides** present in the food chain are harmful as local people hunt food for survival.

Effects of degradation on the environment

Make the Link

Contrast the causes of global warming in the rainforest to the effects of global warming on the tundra.

The tundra is a very fragile environment and the smallest pressures can bring about its destruction. Oil spills have caused severe water pollution in the Arctic Ocean. Burst pipes have spilled vast amounts of crude oil in Alaska devastating the environment and wildlife. Some animals' movements to traditional feeding and nesting grounds have been disrupted by pipelines built above the ground. Pollution from mining and oil drilling has contaminated the air, lakes and rivers. Pesticides used to control swarms of insects have poisoned thousands of **migrating birds** who feed on them. **Polar ice caps** are melting because of an increase in global warming. As a result, many polar bears have starved to death because their feeding patterns have been disrupted. Melting ice caps are causing sea levels to rise and low-lying coastal areas to flood. Global warming is melting the permafrost and every year several metres of tundra are disappearing. As the tundra melts, plant matter that has been frozen for thousands of years is now decomposing. This process is returning carbon dioxide to the atmosphere, increasing global warming. As a result, Alaska is now considered a **carbon source** and not a **carbon sink**.

Management – strategies to minimise impact/effects

Wherever possible, countries exploit the resources found in their tundra environments to help them increase revenue and secure energy supplies. To ensure the tundra is conserved for future generations, **sustainable development** is vital for its survival.

Management strategies designed to do this include:

1. The **Arctic National Wildlife Reserve** in Alaska is the largest conservation area in the Arctic and it has been protected for over 60 years. However, it is under threat from development as huge oil reserves have been discovered by American companies.

2. **Biodiversity Action Plan**. Many tundra areas including Arctic Canada are protected through this internationally recognised program that is designed to protect threatened species and restore habitats.

3. **Habitat Conservation Programs** can be established in tundra environments to protect the unique habitat of tundra wildlife.

4. **Eco-tourism**. This can help to protect certain areas as tourists' money is invested in conservation programs.

5. **Reducing global warming** is critical to protecting the tundra environment as the heating up of Arctic areas is threatening the existence of this fragile environment. Most governments have promised to reduce greenhouse gasses by signing up to the **Kyoto Protocol**.

6. **Renewable energy**. Many countries have invested in renewable energy such as wind, wave and solar power. These energy sources are more **environmentally friendly** than burning **fossil fuels**, which increase global warming.

7. With **energy consumption** rising, it is important that industries, transport companies and **consumers** conserve energy and reduce their **carbon footprint** – the amount of carbon produced. This can be done in many ways such as airlines including a **carbon tax** in the cost of their flights.

8. **Think Global, Act Local Campaign**. It is important that people use energy more efficiently, especially in their homes, so that less is wasted. This can be done in a number of simple and easy ways:

   - Switching off lights, power sockets and TVs when not in use.
   - **Recycling** glass, cardboard and plastic and, reusing bags.
   - Using **energy-efficient light-bulbs** and **rechargeable batteries**.
   - Insulating house roofs and using more efficient heating systems.
   - Installing **solar panels** to house roofs.
   - Walking, cycling, or using **public transport** rather than using cars.

 HINT

Think about ways in which you can help to reduce your carbon footprint.

GO! Activity 2: Paired activity

1. In pairs, choose either the rainforest or the tundra environment.
2. Design and make an environmental campaign poster for Greenpeace.
3. Your poster should include:
   (a) An eye-catching title.
   (b) A list of bullet points describing how your chosen environment is being misused.
   (c) A diagram showing why this environment should be protected.
   (d) An outline of strategies that are designed to protect the environment.
   (e) Pictures and graphs to back up your written points.

GO! Activity 3 (National 5)

Individually, research and write a newspaper report comparing the impact of human activity on the rainforest and tundra environments. You should include:

- A world map showing the location of the rainforest and tundra environments.
- A description of the use and misuse of both environments.
- An explanation comparing the impact of developments and degradation on people and the environment in both areas.
- An outline of any **two** management strategies for the rainforest and tundra.
- **Two** graphs showing data of your choice, e.g. two climate graphs comparing the climate of the rainforest and tundra environments.

Summary

In this chapter you have learned:

- A description of the climates and ecosystems of rainforest and tundra environments.
- How these environments are used and misused by people.
- The effects of degradation on people and these environments.
- Management strategies to minimise the impact and effects of degradation.

You should have developed your skills and be able to:

- Locate rainforest and tundra environments on a map.
- Draw and interpret climate graphs.
- Extract, interpret and present numerical and geographical information which may be statistical, graphical or tabular.

End of chapter questions

National 4 questions

(a) Describe the temperature **and** rainfall in the rainforest.

(b) Name the **four** main layers of rainforest vegetation.

(c) Outline the importance of the rainforest's nutrient cycle.

(d) Name **two** ways the rainforest environment is used.

(e) List **four** different activities that misuse the rainforest environment.

(f) Describe the temperature **and** rainfall in the tundra.

(g) Explain why trees cannot grow in the tundra environment.

(h) Describe the tundra ecosystem in summer **and** winter.

(i) List **four** different activities that misuse the tundra environment.

(j) Describe **two** ways the tundra environment can be managed to minimise the impact of developments.

National 5 questions

(a) Look at Figure 10.3. **Describe, in detail,** the climate of Manaus in Brazil.

(b) **Explain** why the rainforest is a very fragile ecosystem.

(c) **Describe** the ways the rainforest is used **and** misused by different groups of people.

(d) **Explain** the effects of rainforest degradation on people **and** the environment.

(e) **Outline** at least **two** management strategies intended to minimise the effects of rainforest degradation.

(f) Look at Figure 10.11. **Describe, in detail,** the climate of Iqaluit in Nunavut.

(g) **Explain** why the tundra is a very fragile ecosystem.

(h) **Describe** the ways the tundra is used **and** misused by different groups of people.

(i) **Explain** the effects of degradation of the tundra environment on people **and** the environment.

(j) **Outline** at least **two** management strategies intended to minimise the effects of degradation of the tundra environment.

National 5 exam-style questions

You can find sample answers to these exam-style questions on the Leckie website:
https://collins.co.uk/pages/scottish-curriculum-free-resources

1. Study Diagram Q1 below.

   Compare the data shown in the two climate graphs.

   (4 marks)

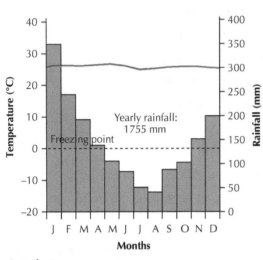

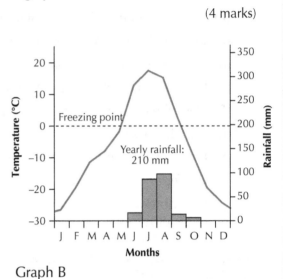

Graph A   Graph B

Diagram Q1: *Climate graphs of selected areas*

2. Study Diagram Q2 below.

   Explain the impact of deforestation on the local people and environment.

   (6 marks)

Diagram Q2: *Logging in Indonesia*

3. Referring to an area you have studied, describe **two** management strategies that help to minimise the effects of rainforest degradation.

   (6 marks)

4. **Describe** the use and misuse of the tundra environment.

   You should refer to an area that you have studied in your answer.

   (6 marks)

Learning checklist

Now that you have finished the **Natural regions** chapter, complete a self-evaluation of your knowledge and skills to assess what you have understood. Use traffic lights to help you make up a revision plan to help you improve in the areas you identified as red or amber.

- Locate the equatorial climate region on a map of the world.

- Draw a climate graph by plotting a line to show temperatures and bars to display rainfall.

- Describe the equatorial climate from a climate graph.

- Describe the equatorial rainforest ecosystem.

- Outline the use of the equatorial rainforest by people.

- Explain the misuse of the equatorial rainforest by people.

- Discuss the effects of degradation on the people.

- Discuss the effects of degradation on the environment.

- Describe the management strategies designed to minimise the impact/effects.

- Locate the tundra climate region on a map of the world.

- Describe the tundra climate from a climate graph.

- Describe the tundra ecosystem.

- Outline the use of the tundra environment by people.

- Explain the misuse of the tundra environment by people.

- Discuss the effects of degradation on the people.

- Discuss the effects of degradation on the environment.

- Describe the management strategies designed to minimise the impact/effects.

Glossary

Aerial photographs: Pictures taken from above that show a birds-eye view of an area.

Afforestation: Planting trees in an area.

Agro-forestry: When crops and trees are grown at the same time.

Annual: Every year.

Arctic Circle: An imaginary line of latitude that circles the globe at latitude 66 degrees north of the Equator.

Arctic National Wildlife Reserve: A large conservation area in the Arctic that has been protected for over 60 years.

Biodiversity Action Plan: A report that outlines aims to protect and conserve plants and animals in an environment.

Buttress roots: A large tree root that grows above the forest floor to provide nutrients and support.

Camouflage: An ability to blend in with a background.

Canopy: The layer of continuous tree tops in the rainforest.

Carbon dioxide sink: A place that absorbs a lot of the gas carbon dioxide.

Carbon footprint: The amount of carbon that each person produces.

Carbon source: Any process that releases carbon into the atmosphere.

Carbon tax: Money charged on the carbon content of fuels.

Caribou: An animal native to the tundra (also called a reindeer).

Cattle ranching: When herds of cattle are farmed on a large area of land.

Chagra: The name for a clearing used by tribes to grow crops.

Climate change: The long-term alteration of weather patterns over time.

Coffee plantations: Rows of shrubs that bear coffee beans.

Cold desert: A natural region that experiences less than 250 mm of precipitation and has cold temperatures throughout the year.

Commercial developments: Human activities that destroy the rainforest to make money.

Consumers: People who are the final users of a product.

Convectional rainfall: When the sun heats the air, warm, moist air rises, cools, condenses and forms clouds. These large, towering clouds burst and heavy rain falls.

Debts: The amount of money that a country has to pay back.

Decaying: When plants wither and die.

Deforestation: The removal of trees from a forest.

Degradation: The deterioration of an area over time.

Desert: A natural region that experiences less than 250 mm of precipitation and has either hot or cold temperatures throughout the year.

Disease: An abnormal condition that affects the body, e.g. malaria.

Ecosystem: A living community of plants and animals sharing an environment with non-living things such as climate, soil and water.

Eco-tourism: When small groups of people visit remote, untouched environments to support conservation and observe natural vegetation and wildlife.

Emergents: The largest trees that grow up through the canopy.

Energy consumption: The total energy used by all human beings.

Energy-efficient light-bulbs: Light-bulbs which use a low amount of energy.

Entertainments: Facilities that are built for people to use in their free time.

Environmentally friendly: Activities that are not damaging to the natural environment.

Equator: An imaginary line of latitude that divides the Earth into a northern hemisphere and a southern hemisphere.

Equatorial climate: The usual weather experienced in rainforest regions – typically warm and wet.

Erosion: The wearing away of the land.

Evaporation: When water changes from a liquid to a gas.

Exploitation: When the natural environment is destroyed for its natural resources.

Exported: When goods are taken from one place to another.

Extinction: When a type of plant or animal is wiped out forever.

Fertiliser: Chemicals which help crops to grow.

Flooding: An overflow of water that submerges land.

Forest floor: The ground in a rainforest.

Fossil fuels: Fuels which are finite, e.g. oil, coal and natural gas.

Future generations: People who will be alive in many years to come.

Global warming: The increase in the earth's temperature over a period of time.

Growing season: The length of time it takes for plants to fully mature.

Habitat: The area where wild plants and animals live.

Habitat Conservation Programs: When the places where animals live are protected.

Humidity: The amount of moisture in the air.

Humus: A layer of dead and decaying leaves found on the forest floor.

Hydro-electric power schemes (HEP): Energy which is generated from fast-flowing water.

Immunity: When a person's body has resistance to a disease.

Industry: The work that people do.

Infrastructure: The framework for allowing people to exist in a place such as roads, houses and schools.

Inuit hunters: Native people who live in the Arctic and stalk and catch animals for food to survive.

Iron ore: A rock containing the metal iron.

Kayapo Amerindian Tribe: A native group of people living in the Amazon Rainforest.

Kyoto Protocol: An agreement by nations to reduce carbon emissions.

Lake: A pool of water on the surface of the landscape.

Leaching: When heavy rainfall washes the goodness out of the soil.

Lichen: A type of fungus that grows on rock.

Logging: The removal of trees for timber.

Machete: A long knife used for clearing land.

Manioc: A crop grown in the rainforest to provide carbohydrates to the diet.

Marsh: An area of land submerged in water.

Medical research: The on-going investigations into cures for diseases.

Medicines: Drugs which are taken to cure illness and disease.

Migrating birds: Birds which fly in and out of an area depending on the time of year.

Minerals: Solid substances located underneath the ground which are mined and sold for profit.

Mining: When natural resources are extracted from underground.

Mosses: Small plants that grow in damp or shaded places.

National park: An area that is protected by law to ensure its conservation.

Native people: Tribes who live in the rainforest.

Natural environment: A place which has not been made by man – it is naturally occurring.

Natural gas: A naturally occurring hydrocarbon gas mixture.

Northern hemisphere: The top half of the earth located north of the Equator.

Nutrient cycle: The continuous movement and exchange of organic matter such as humus back into the production of living matter, e.g. leaves.

Nutrients: Chemical elements which are essential for plant nutrition.

Oil field: An area either on land or under the sea that contains oil.

Oil rig: A large structure built above an oil field to extract oil from underneath the ground or sea.

Permafrost: Permanently frozen soil.

Pesticides: Chemicals which kill pests that attack and eat crops.

Polar ice caps: A large area of land covered in ice.

Pollution: When the air, water or land is damaged by harmful chemicals.

Precipitation: Any type of moisture falling from the sky – rain, hail, sleet and snow.

Profits: The amount of money made after all expenses have been paid.

Public transport: Trains, buses and taxis available for the public to use for a charge.

Rainfall: Water droplets that fall from clouds.

Rechargeable batteries: Batteries that can be used over and over again as they can be repowered.

Recycling: A process of changing waste materials into new products to prevent the waste of useful materials.

Regeneration: Re-growth of vegetation after it has been removed by shifting cultivators.

Renewable energy: Power generated using natural resources that will not run out, e.g. wind and wave power.

Reservations: Designated areas which are protected from developments in the rainforest.

Reuse: To use an item again and again.

Sandpiper: A type of bird that migrates into the tundra during summer months.

Satellite: Equipment that circulates the earth in space and sends back photos.

Seedling: A young plant developing from a seed.

Selective logging: When trees are only cut down once they reach a certain height.

Shifting cultivation: A type of farming that involves people moving every 2–3 years to a new area of the rainforest.

Shrub layer: The area above the ground containing some vegetation.

Shrub: Small bush-like plant.

Slash and burn: A traditional type of farming involving the removal of vegetation with machetes and then setting it on fire.

Solar panel: A piece of equipment used to generate electricity by capturing heat from the sun.

Species: Different varieties of plants and animals.

Standard of living: How well off the people of a country are in terms of wealth, health and education.

Sustainable: When a resource can be used over and over without it being totally destroyed.

Sustainable development: When countries use their resources in a way that ensures their use for future generations.

Sustainable forestry schemes: When trees are cut down and re-planted in a manageable way to ensure the entire forest is not destroyed.

Temperature range: The difference between the highest and lowest temperatures.

Timber: Wood used in building and making furniture.

Traditional lifestyle/ Traditional way of life: The behaviour, habits, ideas and customs which are typical of a particular society.

Trans-Alaskan Oil Pipeline: A very long pipe that transports oil across Alaska.

Trans-Amazonian Highway: A long road which goes right through the Amazon Rainforest.

Tropical hardwood trees: Trees such as teak, mahogany and rosewood which produce high quality timber.

Tropical rainforests: A natural region located between the Tropics.

Tundra: A natural region located above the Arctic Circle.

Under canopy: The area of rainforest vegetation that grows between the canopy of trees and the forest floor (also called the under storey).

Vegetation: Trees and plants.

Vines: Plants which grow down from trees in the rainforest (also called lianas).

Water cycle: The continuous movement of water on, above and below the earth.

Western diseases: Illnesses which are not traditionally found in rainforests, such as measles.

Wildlife: Non-domesticated/wild plants and animals.

World market: Countries across the globe which are able to buy resources.

Yams: A vegetable crop grown in the rainforest.

11 Environmental hazards

Within the context of environmental hazards, you should know and understand:

- Main features of earthquakes, volcanoes and tropical storms.
- Causes of each hazard.
- Impact on the landscape and population of each hazard.
- Management – methods of prediction and planning and strategies adopted in response to environmental hazards.

You also need to develop the following skills:

- Identify on a map the location of earthquakes, volcanoes and tropical storms.
- Extract, interpret and present numerical and geographical information which may be statistical, graphical or tabular.

Environmental hazards

An environmental hazard is a natural event that takes place in a specific area, often without warning, and has a negative effect on people and the landscape. Examples of environmental hazards include **earthquakes**, **volcanoes** and **tropical storms**.

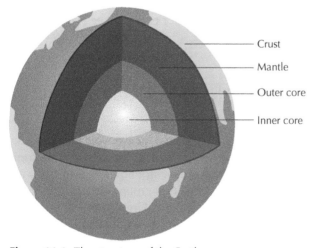

Crust
Mantle
Outer core
Inner core

Figure 11.1: *The structure of the Earth*

To understand why earthquakes and volcanoes occur, it is important to know the structure of the Earth. Figure 11.1 shows the different parts of the Earth. The crust, which is made of solid rock, floats on top of the **mantle**. The mantle is a thick layer of hot **molten rock** called **magma**.

HINT

When magma rises towards the crust it often bursts through a weakness and forms a volcano.

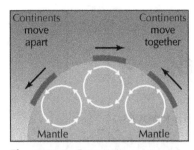

Figure 11.2: *Convection currents in the mantle*

> 🔍 **HINT**
>
> Magma (molten rock) underneath the crust is the consistency of thick jam. Imagine the crustal plates are giant jigsaw pieces floating on top of it.

> 🔍 **HINT**
>
> Fold mountains are formed when tectonic plates are pushed together.

> 🔍 **HINT**
>
> Plates move at about the same rate as your hair and nails grow!

> 🔍 **HINT**
>
> You may know of some volcanoes in your local area, e.g. Arthur's Seat in Edinburgh, that are not located on plate boundaries but these are most likely very old and extinct now.

Figure 11.2 shows convection currents in the mantle. It is these heat currents that enable the **crustal plates** to move. When magma rises towards the crust, the plates are pulled apart, and when magma sinks towards the core, the plates move together.

Plate tectonics

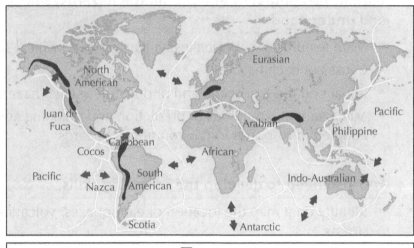

Figure 11.3: *Tectonic plates*

The **Earth's crust** is broken into about 15 giant pieces called **tectonic plates**. Plates are made up of continental or oceanic crust. **Continental crust** is light and very strong but **oceanic crust** is heavier and sinks below continental crust. The convection currents in the mantle cause the plates to constantly move, about 2–3 centimetres every year. Figure 11.3 shows the names of the largest plates and the arrows illustrate the direction they move. Plates can either move side by side, apart or together. The edges where they meet are called **plate boundaries** or **fault lines** and most major earthquakes and volcanoes occur there.

The location of earthquakes and volcanoes

Figure 11.4 shows that earthquakes and active volcanoes mainly occur along fault lines which are located at the edges of tectonic plates. There are thousands of volcanoes around the world. They are confined to specific areas in narrow belts or groups. Around the Pacific Ocean, active volcanoes form the **Ring of Fire**. In the centre of the Atlantic Ocean a line of volcanoes forms the **Mid-Atlantic Ridge** and contains volcanic islands like Iceland and the Canary Islands. Earthquakes can occur anywhere but destructive earth movements are confined to certain areas. The west coasts of North and South America are located in earthquake zones. An area of activity also runs through the Mediterranean Sea around Turkey and Italy where the Eurasian and African plates meet.

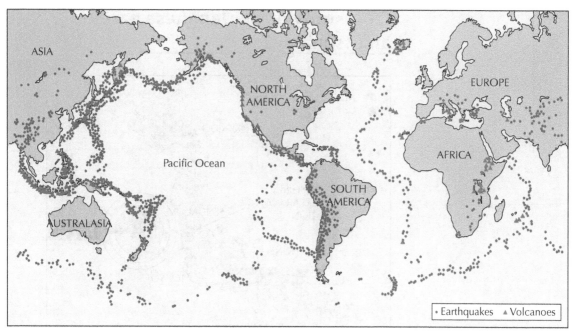

Figure 11.4: *The location of earthquakes and volcanoes*

The main features of earthquakes

An **earthquake** is a sudden and violent shaking of the ground caused by a sudden slip on a fault. A fault line is the boundary between two crustal plates. The **focus** of an earthquake is the point underground at which the plates snap apart. The **epicentre** is the area directly above the focus on the surface of the Earth and it is the strongest part of an earthquake. **Shockwaves** move away from the epicentre deep below the crust and travel away from it and across the Earth's surface.

The causes of earthquakes

Earthquakes are caused by the sudden movement of tectonic plates which form part of the Earth's crust. Tectonic plates normally slide past one another at a **conservative plate boundary**. At times, however, the two plates can become locked together. Tension increases until eventually the pressure becomes too great and the crustal rocks snap apart. This sudden movement of the Earth is known as an earthquake. Figure 11.6 illustrates the movement of the plates at a conservative plate boundary. As well as sliding in different directions, plates can also move in the same direction but at different rates. The best-known conservative plate boundary is the San Andreas Fault in San Francisco.

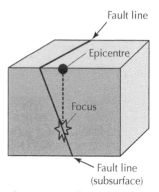

Figure 11.5: *The main features of an earthquake*

🔍 **HINT**

The **fault line** is a **f**racture in the Earth's crust.

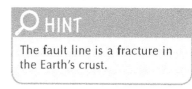

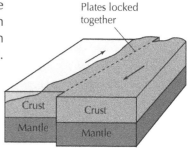

Figure 11.6: *A conservative plate boundary*

Case study of the Japanese earthquake and tsunami 2011

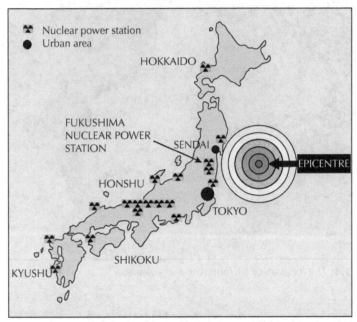

Figure 11.7: *Location of Japanese earthquake and tsunami*

Cause of the Japanese earthquake and tsunami

On Friday, 11 March 2011, a devastating earthquake measuring 9·1 on the **Richter scale** struck eastern Japan. Despite being the fifth strongest earthquake ever recorded, little damage was caused by the initial earthquake and there were few deaths. However, as it occurred under the Pacific Ocean, the earthquake triggered a catastrophic **tsunami**. The focus of the earthquake occurred 80 km east of Honshu, Japan's largest island. The cause of the earthquake was the **subduction** of the Pacific Plate under the Philippine Plate. The crustal rocks wedged as they tried to move past one another. At 2.46 pm local time, the pressure build-up was too great and the crustal rocks snapped, causing an earthquake. As this happened underwater, the ocean floor was pushed upwards, displacing a huge volume of water. This caused a devastating tsunami that spread in all directions across the Pacific Ocean. The huge 10 metre tidal wave rapidly travelled to the coast of Honshu at Sendai where it moved several kilometres inland.

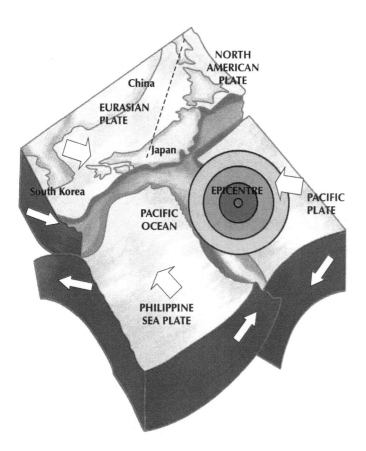

The Great Japanese Earthquake 2011

Figure 11.8: *Plate boundaries affecting Japan*

Impact of earthquake and tsunami on the landscape and population

The impact on the landscape was catastrophic and wiped out thousands of coastal buildings, including entire towns. Over 14 000 houses were totally destroyed and more than 100 000 homes were damaged. Communications were cut off, including phone and electricity lines, leaving more than 256 000 homes without power. Over 1 million people were left without running water and sewage pipes were ruptured. Transport networks, including many roads, were ruined. Fires broke out due to severed gas pipes and serious damage was caused to four out of six Fukushima nuclear reactors. Extensive areas of **farmland** and crops were wrecked. The total cost of the damage was estimated at over US$232 billion.

The effects on people were also disastrous with over 15 000 **fatalities**. More than 6000 people suffered injuries and over 200 000 people were evacuated from their homes. There was a lack of fresh drinking water and food as many fish died and farms were destroyed. More than 215 000 people were displaced to emergency shelters as the tsunami washed away their houses.

Figure 11.9: *Damage caused by the earthquake*

Figure 11.10: *Fukushima nuclear power plant*

Figure 11.11: *Sendai airport was heavily damaged by the tsunami*

> ### 🔍 HINT
>
> Think about the impact of the damage caused by radioactive material leaked from the Fukushima nuclear power plant.

Management – methods of prediction and planning and strategies adopted in response to earthquakes

Methods of prediction

Earthquakes are extremely difficult to predict. Experts know which areas have a high risk of earthquakes and they can identify **frequency patterns** from previous earthquakes. This means that **active earthquake zones** are closely monitored for **seismic activity**. Various instruments are used including **seismographs** (seismometers) which can be used to detect earth movements. Tiltmeters are used to measure changes in the gradient of the crustal plates. Sophisticated sound recording equipment is also used to detect **earth tremors**. These give scientists precious time to issue tsunami warnings in coastal areas.

Figure 11.12: *A seismograph records earth movements*

Methods of planning and strategies adopted in response to earthquakes

Developed countries like the USA and Japan have more money than **developing countries** to invest in specialised equipment, and earthquake prone areas have a number of planning measures in place. Japan is well-prepared for earthquakes as it experiences around 20% of the world's most serious seismic events. Schoolchildren practise **earthquake drills** and householders keep earthquake survival kits. **Earthquake proof buildings** have been constructed to sway with the movement of the earth. Messages are texted to mobiles within seconds of an earthquake to warn people and give advice about getting help. **Emergency services** are better prepared and equipped as they are trained to deal with the aftermath of earthquakes. Rescue teams have specialist heat-seeking equipment and sniffer dogs are trained to locate survivors.

Developing countries like Haiti do not have the money to invest in these measures and that is why the death and destruction caused by earthquakes is often worse. They also lack the communications which allows swift **evacuation**. Many developing countries rely on **international aid** to help them recover from an earthquake as they do not have the resources to recover by themselves. Even although Japan was the second richest country in the world at the time of the earthquake, social networking sites on the internet, e.g. Facebook, helped people donate to disaster relief funds. Other websites helped to locate missing people, e.g. Google Person Finder.

Make the Link

Developing countries often suffer more from the effects of earthquakes. Consider how environmental hazards affect the standard of living of people in poor countries.

🔵 Activity 1: Group activity

1. In groups investigate:

   (a) An earthquake that has occurred in a developed country, e.g. Japan, and an earthquake that has occurred in a developing country, e.g. Haiti.

   (b) The impact of the earthquakes on both countries, e.g. number of deaths.

   (c) The prediction and planning procedures in place.

2. You must include location maps, graphs and photos to illustrate your findings.

3. You should display the information your group has gathered in a poster.

4. Decide who will talk about each section and practise your presentation.

5. Present your group poster to your class.

The main features of volcanoes

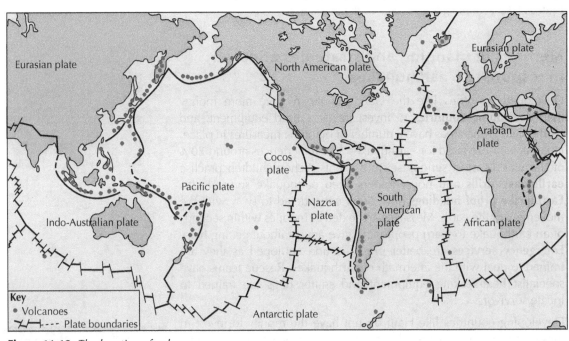

Figure 11.13: *The location of volcanoes*

🔍 **HINT**

Most volcanoes are located at the edges of plates.

Volcanoes are made up of layers of ash and **lava**. The hole at the top of the volcano is called the **crater**. The vent connects the **magma chamber** beneath the crust to the crater. This allows magma to travel up through the crust and out the top of the volcano. As well as lava, ash, gas, dust and lava bombs are emitted from a volcano. If the **main vent** becomes blocked, a **side vent** will form and magma will flow sideways from the volcano. When molten rock reaches the surface of the Earth, it is called a **volcanic eruption**. Magma at the surface is called lava and it cools and hardens into solid rock. This process can repeat itself over many years forming a cone-shaped mountain or volcano.

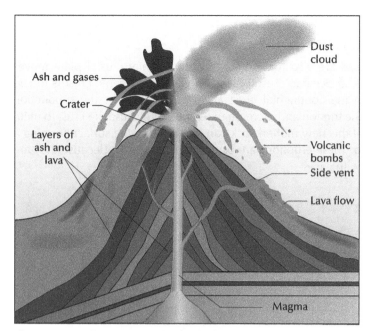

Figure 11.14: *The structure of a volcano*

> **HINT**
>
> A secondary cone can form when lava flows from a side vent.

The causes of volcanoes

The formation of volcanoes at constructive boundaries

Volcanoes can happen in two different ways, depending on the type of plate margin. At a **constructive plate boundary**, two plates move away from each other which allows magma to rise up through the gap in the crust. This helps to form a new oceanic crust found under the ocean. As a result, the Atlantic Ocean is getting bigger by about 3 centimetres every year! The Mid-Atlantic Ridge runs the length of the Atlantic Ocean and surfaces at Iceland – a volcanic island.

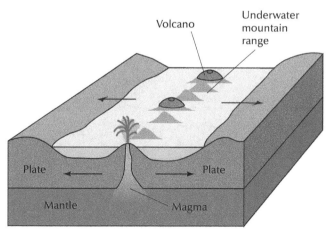

Figure 11.15: *A constructive plate boundary*

The formation of volcanoes at destructive boundaries

Volcanoes are also formed when a continental crustal plate moves towards an oceanic crustal plate. The heavier oceanic crust is forced beneath the continental crust at an area known as the subduction zone. The friction and increase in temperature cause the crust to melt. Some of this new molten rock can be forced to the surface forming a volcano. These eruptions are usually very explosive because they are mixed with gases.

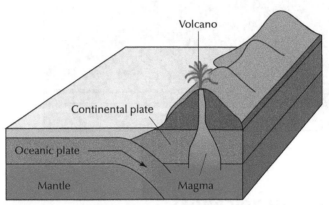

Figure 11.16: *A destructive plate boundary*

Case study of Montserrat

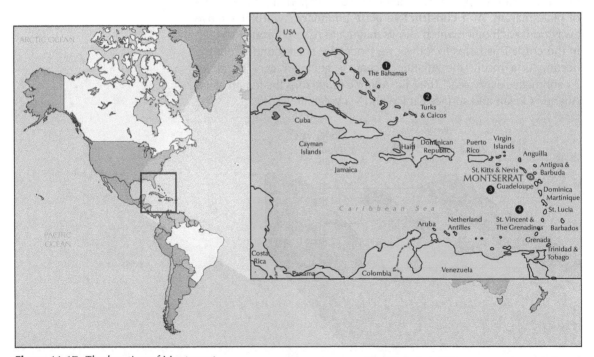

Figure 11.17: *The location of Montserrat*

Cause of the Montserrat volcanic eruption

On 18 July 1995, the Soufrière Hills volcano on the Caribbean island of Montserrat became active after 400 years of inactivity. Montserrat lies on the boundary between the Atlantic Plate and the Caribbean Plate. The **destructive plate boundary** meant that the oceanic crust was forced beneath the continental crust. This movement triggered an explosive volcanic eruption. Friction and heat caused the plate to melt, forming magma. The molten rock built up until it had the chance to reach the surface through cracks in the Earth's crust. Since the initial eruption, there has been on-going seismic activity and eruptions.

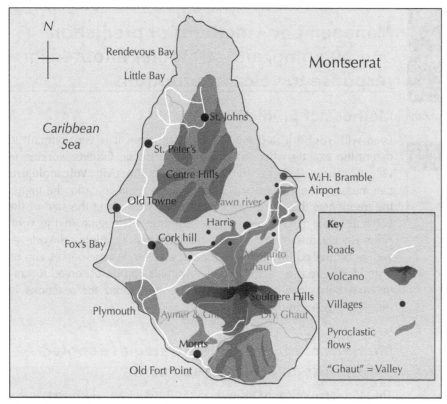

Figure 11.18: *Montserrat Island post-eruption*

Impact of the volcano on the population and landscape

In April 1996, authorities set up an **exclusion zone** and ordered people to leave the capital city, Plymouth, and south of the island. As volcanic activity increased, more than half the island's population was evacuated to other Caribbean Islands and the UK. A voluntary evacuation scheme was introduced to allow evacuees to temporarily move to the UK for two years. On 25 June 1997, the huge volcanic dome collapsed spewing five million cubic metres of lava and ash in a devastating **pyroclastic flow**. As a result, 19 people were killed.

Figure 11.19: *The Soufrière Hills volcano*

Figure 11.20: *Destruction of Plymouth*

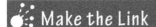

Make the Link

Consider the positive and negative impact of tourism on the island of Montserrat.

The eruption destroyed many towns and villages including the capital city. At least 4 square kilometres of land were covered in lava. Sixty per cent of the south and central part of the island remains uninhabitable. Transport networks including roads and the main airport on the eastern side of the island were destroyed. Crops were ruined in the fertile fields and fish were poisoned in the waters surrounding the island. Since early 2010, the volcano has been relatively quiet but it continues to be closely monitored by the Montserrat Volcano Observatory. The hills continue to smoulder and the people who remain on the island are making money from tourists visiting the active volcano.

Management – methods of prediction and planning, and strategies adopted in response to volcanic eruptions

Methods of prediction

Even with sophisticated monitoring equipment it is very difficult to determine exactly when an eruption will occur. Experts working in volcano observatories can try to predict volcanic activity. **Volcanologists** can measure the frequency of earthquakes on the volcano: the higher the frequency, the closer the eruption. By measuring the size of the **volcanic cone** it can confirm the build-up of magma in the vent. Scientists can also check for **gas emissions**, mainly sulphur dioxide, as well as increased **thermal activity** at the crater. Water sources can be tested for increasing acidity levels. Animals can be observed for any unusual behaviour and plants can be monitored for reactions to increasing gases in the atmosphere.

Methods of planning and strategies adopted in response to volcanic eruptions

The Montserrat Volcano Observatory (MVO) devised a risk management zone map. The military developed **evacuation plans** and the authorities were able to evacuate people from the areas surrounding the volcano. An exclusion zone was set up around the volcano. **Emergency sirens** sounded to warn people to leave the area when seismic activity alerted scientists to an imminent eruption. Emergency services were on standby and ready to assist people in need of help. The UK government agreed an aid package to help the government of Montserrat recover from the aftermath.

The location of tropical storms

NORTH
AMERICA

ASIA

EUROPE

Tropic of
Cancer 23½°

Typhoons 26/yr
(May–Dec)

Equator 0°

Hurricanes
13/yr (June–Oct)

Hurricanes
9/yr (Aug–Oct)

AFRICA

Cyclones
6/yr (Oct–Nov)

Typhoons
10/yr
(Jan–Mar)

Tropic of
Capricorn 23½°

SOUTH
AMERICA

Cyclones
8/yr (Dec–Mar)

OCEANIA

10/yr Average number per year
⬛ Warm sea areas (over 27°C)   ☐ Cooler sea areas   ▶ General direction of storms

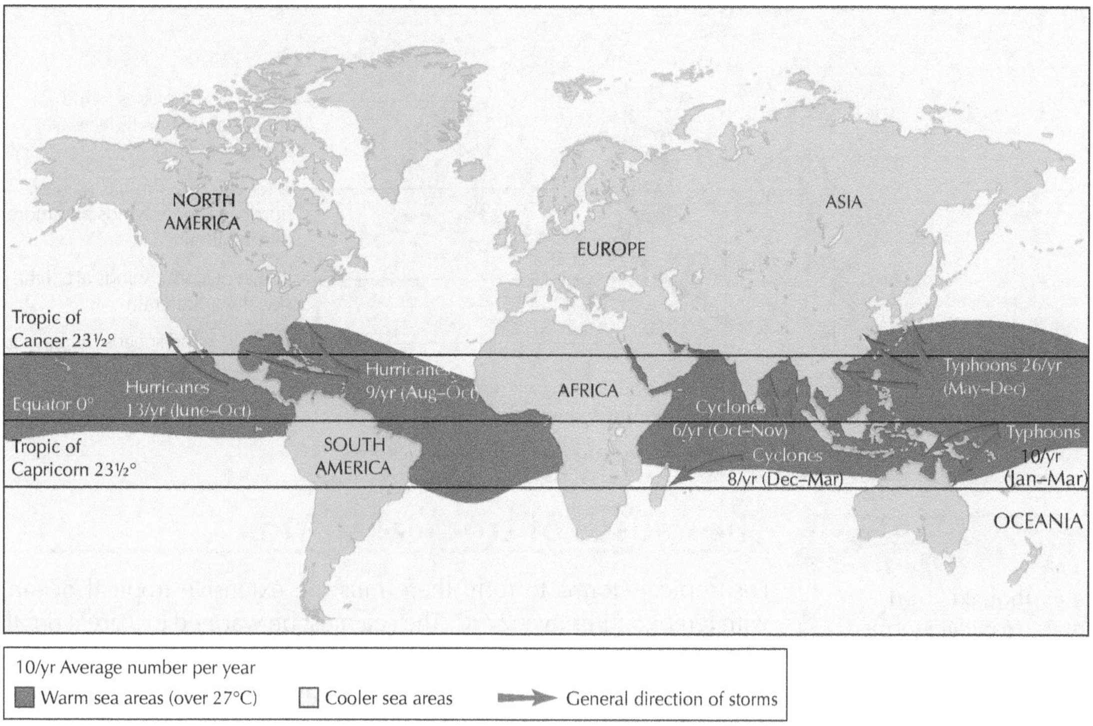

Figure 11.21: *Global distribution of tropical storms*

Tropical storms are found between the **Equator** and the **Tropics**. They originate over the eastern side of oceans and move westwards where they cross land causing immense destruction before dying out. Tropical storms are known locally by different names. They are called **hurricanes** in the USA, **cyclones** in India and **typhoons** in China and Japan.

The main features of tropical storms

A tropical storm is a rapidly revolving storm system. It is characterised by low **air pressure** which creates very unstable conditions including thick clouds, strong winds and a spiral arrangement of thunderstorms which bring torrential rainfall. The **eye of the storm** is located in the centre of the spiralling air where conditions are typically calmer and drier.

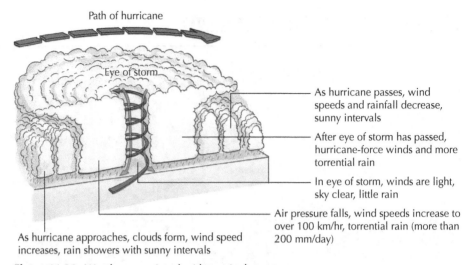

Path of hurricane

Eye of storm

As hurricane passes, wind speeds and rainfall decrease, sunny intervals

After eye of storm has passed, hurricane-force winds and more torrential rain

In eye of storm, winds are light, sky clear, little rain

Air pressure falls, wind speeds increase to over 100 km/hr, torrential rain (more than 200 mm/day)

As hurricane approaches, clouds form, wind speed increases, rain showers with sunny intervals

Figure 11.22: *Weather associated with tropical storms*

The causes of tropical storms

For tropical storms to form there must be extensive tropical oceans with temperatures over 27°C. The sea must be warmed to a great depth at the time of year when water temperatures are at their highest. This results in very moist air spiralling upwards, creating low air pressure. As the air rises, it cools and condenses which releases large amounts of heat. This heat powers the tropical storm.

Case study of Hurricane Sandy

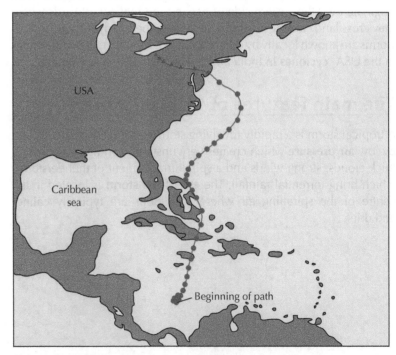

USA

Caribbean sea

Beginning of path

Figure 11.23: *The path of Hurricane Sandy*

Cause of Hurricane Sandy

Hurricane Sandy developed in the Caribbean Sea on 22 October 2012. The air above the sea began to rise and created an area of low pressure. Air was sucked in over the sea to replace the rising air and a **tropical depression** was formed. Winds became stronger as air rose more rapidly, as did the movement of air to replace it. As a result, air started to spiral and increase in speed. The storm quickly strengthened and intensified and on 24 October, Sandy became a hurricane. The hurricane travelled over the Caribbean Sea hitting Jamaica and Cuba. Early on 26 October Sandy moved through the Bahamas. On 29 October, Sandy curved NNW and made landfall near Brigantine in New Jersey. The severe and extensive damage caused by the tropical storm and its unusual merging with a frontal system, led the media and several government agencies to nickname the hurricane 'Superstorm Sandy'.

Figure 11.24: *Satellite image of Hurricane Sandy*

Impact of the hurricane on the landscape and population

Figure 11.25: *Storm damage on the coastline of New Jersey*

The effects of the hurricane were felt on many Caribbean islands and down the east coast of the USA. In New York, a storm surge caused severe **flooding** to streets, tunnels and subway lines and damaged coastal buildings. Power was cut in and around the city and many homes were left without electricity for weeks. Bridges collapsed and roads were damaged. Storm damage in the USA alone was estimated at over US$71 billion. In Jamaica, winds left 70% of residents without electricity and lifted roofs off buildings. In Cuba, coastal flooding and wind damage destroyed 15 000 homes. Valuable farmland and crops were ruined in most countries and caused food shortages in Haiti. **Economic development** was badly affected especially in poorer countries and national **debts** increased due to the huge cost of the repairs.

Figure 11.26: *Damage caused to property in New York*

More than 50 million people were affected on the eastern seaboard of the USA. At least 285 people lost their lives along the path of the storm in seven different countries. In Puerto Rico, one man was swept away by a swollen river. Many people were made **homeless** as their houses were destroyed, e.g. in Haiti about 200 000 people lost their houses. Temporary accommodation was provided by those countries that could afford it, e.g. the USA. People were moved inland and away from the path of the hurricane. Businesses and schools were closed and transport services were suspended.

Management – methods of prediction and planning, and strategies adopted in response to tropical storms

Methods of prediction

Advanced weather equipment enables experts to gather detailed information about tropical storms. **Satellites** circulate the Earth and take photos of cloud patterns which are sent back to computers and show the location of tropical storms. Detailed weather information from **weather stations**, **weather ships** and aircraft help experts to try to predict where the storm will hit. However, hurricanes can change direction frequently and suddenly so it can be very difficult to predict exactly where and when they will make landfall.

> **Make the Link**
>
> Tropical storms are easier to predict than earthquakes because their development and movement can be tracked by satellites.

Methods of planning and strategies adopted in response to tropical storms

The governments of the affected countries issued a tropical storm watch. Storm warnings were given by **meteorologists** to give people time to safeguard their homes and evacuate coastal parts to inland areas. Due to advanced technology and communication systems, people had enough time to board up windows and lay sand bags in front of their property. Evacuation plans were put into action by the countries who could afford them. These helped authorities to clear areas expected to be affected in good time. In many regions, emergency shelters were opened and TV campaigns broadcast details of where people could get help. Flights were cancelled and travel alerts broadcast on TV, internet and radio. Emergency services were kept on stand by and helicopters were drafted in to airlift people from along the coast and offshore islands. The National Guard and US Air Force were put on standby to help with the rescue effort and aftermath of the storm.

> **GO! Activity 2**
>
> 1. Individually, choose an earthquake, volcano or tropical storm.
> 2. Design and write a newspaper report on your chosen environmental hazard.
> 3. Your report should include:
>   - An authentic newspaper name.
>   - An eye-catching headline.
>   - Information on where, when, why and how the hazard happened.
>   - A location map of the area affected by the hazard.
>   - An eyewitness interview.
>   - Pictures and graphs to illustrate your findings.

> **HINT**
>
> Google maps is a good resource for getting maps.

GO! Activity 3 (National 5)

Individually, research and write a blog appealing for help intended for an area that has recently suffered from an environmental hazard. You should include:

- A world map showing the location of the area affected.
- A description of the damage caused.
- An explanation of what needs to be done to help people and repair the landscape.
- An outline of any **two** strategies that we can do to help the area affected.
- **Two** graphs showing data of your choice, e.g. bar graphs showing the total cost of damage compared to the cost for repairs.

Summary

In this chapter you have learned:

- Main causes and features of earthquakes, volcanoes and tropical storms.
- The impact on the landscape and population of earthquakes, volcanoes and tropical storms.
- Methods of prediction and planning for and strategies adopted in response to earthquakes, volcanoes and tropical storms.

You should have developed your skills and be able to:

- Identify on a map the location of earthquakes, volcanoes and tropical storms.
- Extract, interpret and present numerical and geographical information which may be statistical, graphical or tabular.

End of chapter questions

National 4 questions

(a) What is an environmental hazard?

(b) What is the name of molten rock found underneath the Earth's crust?

(c) Where do volcanoes and earthquakes occur?

(d) What happens at a conservative plate boundary?

(e) Name **three** instruments used to record earth movements?

(f) What **two** things make up a volcanic cone?

(g) Describe the formation of volcanoes at constructive plate boundaries.

(h) Where are tropical storms located?

(i) Describe the cause of tropical storms.

(j) How do authorities prepare for tropical storms?

National 5 questions

(a) Look at Figure 11.4. **Describe** the location of earthquakes.

(b) Referring to an earthquake you have studied, **explain** how it was caused.

(c) **Describe, in detail,** the impact of an earthquake on the landscape.

(d) **Outline** how experts can prepare for earthquakes.

(e) **Explain two** different ways volcanic eruptions can happen.

(f) Referring to a volcanic eruption that you have studied, **describe** the impact on the local people.

(g) **Explain two** methods of predicting volcanic eruptions **and** comment on their usefulness.

(h) **Outline** the location of tropical storms.

(i) **Explain** the conditions necessary for tropical storms to occur.

(j) **Describe, in detail,** the impact of a tropical storm on the local people and the landscape.
NB: you must refer to a tropical storm that you have studied.

National 5 exam-style questions

You can find sample answers to these exam-style questions on the Leckie website:
https://collins.co.uk/pages/scottish-curriculum-free-resources

1

Diagram Q1: *Structure of a volcano*

Study Diagram Q1 above.

Describe the main features of a volcano. (4 marks)

2 **Describe** the effects of earthquakes on the landscape **and** local people.

You should refer to a specific earthquake that you have studied.

(6 marks)

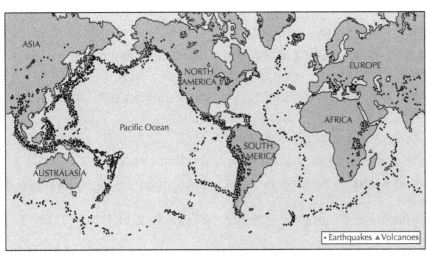

Diagram Q3: *Location of earthquakes and volcanoes*

Study Diagram Q3 above.

3 **Describe** the location of earthquakes and volcanoes. (4 marks)

4 **Explain** how experts can predict **and** plan for an environmental hazard of your choice.

Choose from: earthquakes, volcanoes or tropical storms.

(6 marks)

LEARNING CHECKLIST

Now that you have finished the **Environmental hazards** chapter, complete a self-evaluation of your knowledge and skills to assess what you have understood. Use traffic lights to help you make up a revision plan to help you improve in the areas you identified as red or amber.

- Locate different earthquakes on a world map.

- Describe the features of earthquakes including:

  ➤ fault line

  ➤ shockwaves

  ➤ focus

  ➤ epicentre

- Outline the main causes of earthquakes.

- Describe the impact of earthquakes on the landscape and population.

- Outline the methods of prediction and planning, and strategies adopted in response to earthquakes.

- Locate different volcanoes on a world map.

- Describe the features of volcanoes including:

  ➤ cone

  ➤ crater

  ➤ vent

  ➤ magma chamber

  ➤ side vent

- Outline the main causes of volcanoes.

- Describe the impact of volcanoes on the landscape and population.

- Outline the methods of prediction and planning, and strategies adopted in response to volcanic eruptions.

- Identify the location of tropical storms.

- Describe the features of tropical storms including:

  ➤ eye of storm

  ➤ low air pressure

  ➤ thunderstorms

- Describe the causes of tropical storms including:

  ➤ water temperature

  ➤ location

  ➤ weather patterns

- Describe the impact of tropical storms on the landscape and population.

- Outline the methods of prediction and planning, and strategies adopted in response to tropical storms.

Glossary

Active earthquake zones: Areas where there are frequent earth movements.

Air pressure: The weight of the air in the atmosphere.

Ash cloud: Fine fragments of rock, minerals and volcanic glass created during eruptions and thrown into the atmosphere.

Conservative plate boundary: The area where crustal plates move side-by-side.

Constructive plate boundary: The area where crustal plates move apart.

Continental crust: The layer of rocks which forms the continents.

Crater: The hole at the top of a volcano where lava and ash escape.

Crustal plates: Large pieces of the Earth's crust (also called tectonic plates).

Cyclone: The name for a tropical storm in India.

Debt: The amount of money that a country has to pay back.

Destructive plate boundary: The area where crustal plates move together and one plate is forced beneath the other.

Developing country: A country where people have a low standard of living.

Earth tremors: Movements of the Earth's crustal plates.

Earthquake drills: When people practise what to do in the event of an earthquake.

Earthquake proof buildings: Buildings that are designed and built to withstand an earthquake.

Earthquake: The sudden movement of the Earth's crust.

Earth's crust: The outer shell of the surface of the planet.

Economic development: The increased standard of living and advance in the wealth in a country.

Emergency services: Organisations which ensure the health and safety of the public by addressing different emergencies, e.g. fire brigade.

Emergency sirens: Alarms that sound in the event of an earthquake.

Epicentre: The central part of an earthquake located on the Earth's surface directly above the focus.

Equator: An imaginary line of latitude that divides the Earth into a northern hemisphere and a southern hemisphere.

Evacuation: The removal of people from an area to protect them.

Evacuation plan: A report in place to tell people what to do and where to go if a natural hazard occurs in an area.

Exclusion zone: An area where people are not permitted.

Eye of storm: The centre of a tropical storm.

Farmland: Land which is used to grow crops and/or keep animals.

Fatalities: Deaths.

Fault line: Where one plate meets another (also called a plate boundary).

Flooding: An overflow of water that submerges land.

Floodwater: Water overflowing, often as a result of too much rainfall.

Focus: The point underground directly beneath the epicentre where the crustal rocks snap apart.

Frequency patterns: How often and for how long an earthquake occurs.

Gas emissions: Small particles released from a volcano.

Homeless: When people have nowhere to stay.

Hurricane: The name for a tropical storm in the USA.

International aid: When one country gives help to another country.

Lava: Molten rock that comes out of a volcano.

Lava flow: Molten rock flowing down the side of a volcano.

Magma : Molten rock found beneath the Earth's crust.

Magma chamber: The area beneath a volcano containing molten rock (magma).

Main vent: The thin narrow pipe in the middle of a volcano.

Mantle: The layer of hot, molten rock located beneath the Earth's crust.

Meteorologist: A scientist who studies weather.

Mid-Atlantic Ridge: A line of volcanoes in the centre of the Atlantic Ocean.

Molten rock: Liquid rock melted by the Earth's heat.

Natural hazard: The threat of a naturally occurring event happening, often without warning, which has negative impacts on people and the landscape, e.g. earthquakes.

Oceanic crust: The layer of rocks which forms the ocean floor.

Pacific Ring of Fire: An area of active volcanoes located around the Pacific Ocean.

Plate boundary: Where one plate meets another (also called a fault line).

Prediction: The ability to determine if an environmental hazard is likely to occur.

Pyroclastic flow: A fast moving cloud of hot gas and rock.

Richter scale: A table which details the strength of an earthquake ranging from 1–12 .

Satellites: Equipment that circles the Earth and sends back photos.

Secondary cone: A smaller volcano that forms on the side of the main volcano.

Seismic activity: When there are earth movements.

Seismograph: An instrument used to measure the strength of an earthquake (also called a seismometer).

Shockwaves: Smaller earth tremors normally experienced after a major earthquake (also called seismic waves).

Side vent: The thin narrow pipe at the side of a volcano formed when the main vent is blocked (also called a secondary vent).

Subduction: When an oceanic crustal plate is forced beneath a continental crustal plate.

Tectonic plates: Huge pieces of rock that fit together to form the Earth's crust.

Thermal activity: Increased heat in a volcanic area.

Tiltmeters: Equipment that is used to measure very small changes in the size of a volcano.

Tropic of Cancer: An important line of latitude located at 23·5 degrees north of the Equator.

Tropic of Capricorn: An important line of latitude located at 23·5 degrees south of the Equator.

Tropical depression: An extreme type of weather characterised by low air pressure, very strong winds and heavy rain (also called a tropical storm).

Tsunami: A tidal wave caused by an underwater earthquake.

Typhoon: The name for a tropical storm in China and Japan.

Volcanic bombs: Large blocks of hot rock thrown from a volcano.

Volcanic cone: The shape of a volcano.

Volcanic eruption: When a mountain with a hole in the top expels lava, ash, gas and other materials.

Volcano: A cone-shaped mountain with a hole (crater) at the top.

Volcanologists: A scientist who studies volcanoes.

Weather ship: A large vessel located at sea that is used to observe and record weather conditions.

Weather station: An observation post on land used to observe and record weather conditions.

12 Tourism

Within the context of tourism, you should know and understand:

- Mass tourism and eco-tourism.
- Causes of/reasons for mass tourism and eco-tourism.
- Impact of mass tourism and eco-tourism on people and the environment.
- Strategies adopted to manage tourism.

You also need to develop the following skills:

- Identify on a map the location of popular tourist and eco-tourist destinations.
- Extract, interpret and present numerical and geographical information which may be statistical, graphical or tabular.

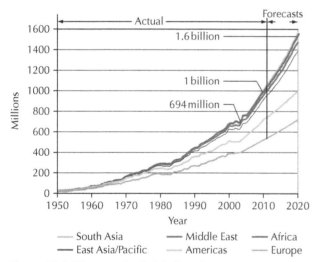

Figure 12.1: *The growth of global tourism*

> 🔍 **HINT**
>
> Compare the growth of global tourism in Europe and South Asia.

Mass tourism

Mass tourism occurs when large amounts of people visit the same place at the same time. It involves large-scale tourism and it is typically related to sun, sea and sand holidays. Mass tourism generates a huge income for foreign tour companies and normally has minimal economic benefits to local destination communities. Mass tourism is a popular form of tourism as it is often the cheapest way to go on holiday abroad. **Package deals** booked through **travel agents** make it low-cost and easy for holiday makers to travel to popular holiday destinations. Examples of mass tourism include:

- Sunbathing on a beach, e.g. the Costa Brava in Spain.
- Skiing in the mountains, e.g. Austria.
- Sightseeing historical ruins such as Pompeii.
- City breaks, e.g. London, New York or Paris.
- Cruise ships sailing to various destinations in one trip, e.g. the Caribbean islands.

HINT

Think about the places you have visited on holiday and the numbers of people that were there.

Figure 12.2: *Mass tourism in New York*

Make the Link

Mass tourism also affects the Yorkshire Dales, the Dorset coast and the Cairngorms in the UK.

Causes of mass tourism

Figure 12.1 shows that tourism has increased since the 1950s. Improvements in road, rail and air travel enable people to travel. Holiday pay and increased time off work gives people the opportunity to visit different places. The impact of **tour operators** and travel agents make it easier to go on holiday due to package deals which often include flights, transfers, meals and **holiday reps** on hand to solve any problems. Cheap package holidays and **budget airlines** make holidays more affordable to many people. Travel programmes and adverts on **social network sites** inspire people to visit different places. The demand to explore various places of interest has increased as **globalisation** has made the world smaller. People also want to experience different climates and guaranteed sunshine.

HINT

Think about where, when, why and how you go on holiday.

Impact of mass tourism on people

Local people are often employed to build **tourist facilities** and work in resorts, e.g. hotels, restaurants and shops. Residents can enjoy improved facilities like theme parks and improvements in **infrastructure** such as electricity, roads and airports. Increased employment means more tax contributions so the local authorities can invest in services such as schools and health care. People can experience different languages and cultures which increases cultural understanding. On the other hand, jobs are often **seasonal**, especially in beach and **ski resorts**, so some people may have no income for several months. Tourists can be noisy and upset the peace and quiet for local residents. Tourist facilities can often be too expensive for local people to use. Tourists can **conflict** with local people as visitors may not respect local cultures. Crime increases as tourists are targeted by thieves.

Impact of mass tourism on the environment

Modern tourist facilities can sometimes improve the appearance of certain places. Some tourists are more environmentally aware and can have a positive impact on the landscape by donating money to local projects, e.g. **nature reserves**. Coastal areas can be cleaned up through initiatives like the **European Union's Blue Flag Scheme** to make them safe for tourists to use. Seas become less polluted as more **sewage treatment** plants are built to reduce water **pollution**. On the other hand, many tourist facilities spoil the look of the natural environment. Farmland and traditional landscapes are often over-developed with large tourist complexes. Increased air travel to foreign resorts contributes to **global warming**. **Traffic congestion** increases air and noise pollution. Increased sewage from visitors increases water pollution. Polluted water damages fish and visitors' litter spoils the appearance of places. Many activities erode the landscape, e.g. skiing and walking. Historical places can become artificial with tourist signs, e.g. Machu Picchu in Peru. Rocks in cultural attractions like Ephesus in Turkey are being worn away by the enormous number of visitors.

> ### Make the Link
>
> You will have studied the impact of mass tourism on at least one National Park in the UK.

Figure 12.3: *Mass tourism on a Spanish beach*

Strategies adopted to manage tourism

Governments, conservation agencies and tour companies all have a responsibility to manage tourism. Strategies vary depending on the tourist area or visitor attraction but many of these strategies will apply to all tourist areas:

- Tourist information centres offer free advice, maps and leaflets to visitors.
- Guides, holiday reps, rangers and tourist police help to educate people on **responsible tourism**.
- Signs in tourist areas give guidance to visitors.
- Designated footpaths direct people to certain areas and away from fragile areas.
- Surfaced car parks keep vehicles off roads and limit congestion on streets.
- Nature reserves and bird sanctuaries help to protect wildlife.
- **The National Trust** and tourist boards promote responsible tourism in the UK.
- Important landscapes and **habitats** can be protected by various measures, e.g. World Heritage Site status and Sites of Special Scientific Interest.
- Byelaws are passed by local councils to protect areas, e.g. wild camping is forbidden on the banks of Loch Lomond.
- A carbon tax is levied on flights to off-set carbon emissions.

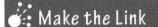

HINT

Tourism generates a lot of income for a country's economy.

Make the Link

Tourist income can be used to help develop a country e.g. improve health care.

Country	International tourist arrivals (millions)	International tourism receipts (US$ billion)
France	82.6	42.5
USA	75.6	205.9
Spain	75.6	60.3
China	59.3	44.4
Italy	52.4	40.2
UK	35.8	39.6
Germany	35.6	37.4
Mexico	35.0	19.6
Thailand	32.6	49.9

Source: United Nations World Tourism Organization

Figure 12.4: *The world's top tourist destinations*

Mass tourism in Spain

Figure 12.5: *Location map of Spain*

Since the 1960s, Spain has been an international tourist destination, earning the country US$60.3 billion per year. Figure 12.4 shows that Spain is the world's third most popular tourist destination, receiving 75.6 million visitors per year. The impact of mass tourism has been dramatic, especially along coastal areas. Traditional fishing and farming villages have been transformed into bustling tourist resorts. Large hotel complexes, restaurants, shops and bars spoil the appearance of the natural landscape and tall buildings block the views of the hills behind the coastline. Traffic congestion is a major problem with hire cars and mopeds blocking roads. Litter left on beaches such as the Costa Blanca is unsightly and harmful to wildlife. The volume of sewage dumped into the Mediterranean Sea is huge due to the number of tourists. Infrastructure, e.g. sewage treatment plants, has struggled to keep pace with the volume of visitors and raw sewage is dumped into the sea. This leads to diseases from people eating contaminated fish or from swimming in the sea.

Tour operators and their holiday reps endeavour to promote responsible tourism. This helps to increase awareness of being respectful towards the local people and wildlife. The European Union's Blue Flag Scheme recognises coastal areas that meet strict criteria including:

- High water quality, e.g. no sewage, industrial or waste water should affect the beach. Guidelines state that sewage must be treated and dumped far out at sea.

⌕ HINT

Consider the reasons why the number of visitors to Spain has decreased from 60 million in 2007 to 52 million in 2012 and then increased to over 70 million in 2016.

- Signs displaying environmental education and information, e.g. a map of **ecosystems** and a code of conduct for responsible tourism.
- Environmental management, e.g. bins and toilets, must be provided.
- Safety and services, e.g. lifeguards and a first aid kit must be available to beach users.

Figure 12.6: *A blue flag in Spain*

Activity 1: Paired activity

1. In pairs, choose a popular tourist area, e.g. Tenerife, Disney World or Las Vegas.
2. Design and write a **brochure advert** on your chosen location.
3. Your advert should include:
   - An eye-catching title.
   - A location map.
   - Climate graph.
   - Information on excursions and places to visit.
   - Details of how to get there, i.e. flights and transfers.
   - Pictures and graphs to illustrate the information.

Activity 2 (National 5)

Individually, research a place where mass tourism occurs. Choose either a **poster** or a **mind map** to display your findings. You should include:

- A definition of mass tourism.
- A location map highlighting your chosen location where mass tourism occurs.
- The key features of mass tourism.
- The main reasons for mass tourism.
- The impact of mass tourism on the local people, the landscape and wildlife.
- **Two** graphs showing data of your choice, e.g. bar graphs showing the income generated from tourism in your chosen place or a climate graph.

Eco-tourism

Eco-tourism is a **sustainable** and **environmentally friendly** form of tourism that involves excursions to fragile and unspoilt areas, including remote areas of rainforest, tundra and hot deserts. Eco-tourism is designed to conserve the natural environment because organised tours are intended to have a low impact and tourist developments are small scale. Visitor facilities are built to blend in with the natural environment, food is obtained locally and waste is usually managed on-site. Tourist numbers are limited to ensure minimal disturbance to the environment. Examples of eco-tourist trips include:

- Organised **wildlife** safaris in Africa's National Parks, e.g. Hell's Gate National Park in Kenya.
- Treks in remote rainforest such as Costa Rica.
- Whale watching in Alaska.
- Home stays with rice farmers, e.g. Thailand.
- Organised tours to ancient sites such as Machu Picchu in Peru.
- Camel trekking in the Sahara Desert.

Reasons for eco-tourism

Many **developing countries** view eco-tourism as an income generator and want to conserve their remote and unusual environments so they can profit from tourism. Some developed countries want to help developing nations conserve their fragile environments by promoting sustainable tourism excursions. Many travellers are bored with package holidays and want to experience a different type of holiday. Some tourists are also more environmentally aware and want to help protect fragile environments.

Figure 12.7: *Eco-tourist river tour in Borneo*

> ## HINT
>
> A key concept of eco-tourism is sustainability which enables forthcoming generations to experience pristine parts of the world in the future.

> ## HINT
>
> Mass tourism is also a cause of eco-tourism as some travellers do not want to be around thousands of other tourists in the same place.

Impact of eco-tourism on people

Eco-tourism provides local people with jobs, e.g. **park rangers**. The money earned is often more than people would make from selling crops which improves their **standard of living**. The education of locals is enriched as they are trained to speak English to give guided tours to visitors. Local communities can also earn money from the sale of handicrafts to tourists. Eco-tourist excursions educate visitors about local issues and encourage them to donate money to ensure the sustainability of local projects. Eco-tourism promotes respect for local cultures and increases global awareness for the **conservation** of native tribes and wildlife under threat from extinction. However, local people can be exploited as **tourist attractions** or cheap labour. Natives lose their **traditional way of life** as they are tainted with 'western' cultures. Many companies running eco-tourist trips make a profit, while local people earn very little money. Infrastructure such as airports is often only beneficial to tourists as local people cannot afford to use it.

Impact of eco-tourism on the environment

Eco-tourism is designed to cause minimal disturbance to natural environments. Destruction of pristine environments is discouraged as it is damaging to income from tourists. **Eco-tourists** are keen to pay for conservation through donations and payments to places of interest. Eco-tourism can reduce the need to hunt animals for income: in West Africa, former poachers are hired as park rangers as they have a detailed knowledge of local wildlife. With eco-tourism, income is earned from environmental **stewardship** and protecting wildlife. On the other hand, eco-tourism is now a large industry in some places like Costa Rica and there are not enough regulations to control companies or people. The damage caused to local resources is a big problem, such as trees being used to make eco-lodges for visitors. Natural resources are also destroyed to make **souvenirs** and animals are captured to use as tourist attractions. Overused trails damage **vegetation** and increase erosion. Pollution is a problem when waste is not properly managed, e.g. waste dumped in rivers contaminating water.

Make the Link

Consider the impact of modern developments in the tundra and rainforest and how these go against the principle of sustainability, limiting the opportunity for eco-tourism in certain areas.

Eco-tourism in Borneo

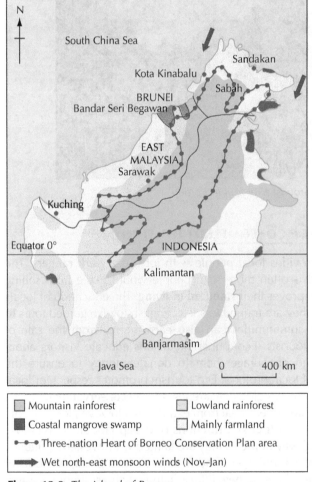

Figure 12.8: *The island of Borneo*

Figure 12.9: *An eco-tourist lodge in Borneo*

Borneo is the third largest island in the world and its location on the Equator means its natural environment is mainly rainforest. However, much of the primary rainforest has been devastated and the native orang-utan is under threat from extinction. The rainforest has been destroyed for many reasons including the sale of tropical hardwood timber and to make way for oil palm plantations. Borneo is the world's largest producer of oil palm which is used in the manufacture of cosmetics, cooking oil and biodiesel fuel for global industries.

National Parks have been set up to protect some of the small areas of rainforest that remain. The Three-Nation Heart of Borneo Conservation Plan was set up by Brunei, Malaysia and Indonesia. The plan aims to protect 250 000 square kilometres of pristine rainforest. Within the area, commercial developments are not permitted but eco-tourist activities which support the environment are allowed to take place. Eco-tourism is now an important industry in Borneo as it generates a lot of income and helps preserve parts of the rainforest. Sepilok Nature Resort is an eco-tourist development in Sandakan. Tourists visit the Orang-utan Rehabilitation Centre to sponsor an orang-utan and ensure its sustainability.

Figure 12.10: *A baby orang-utan*

Make the Link

In the Rural chapter, we learned about the production of biofuels. Consider the impact on Borneo's rainforest and wildlife of producing biodiesel fuel.

Strategies adopted to manage eco-tourism

Governments, conservation agencies and tour companies all have a responsibility to manage eco-tourism. Strategies vary depending on the area but these strategies will apply to many areas:

- National Parks are designated to protect fragile environments from large-scale commercial developments so that profit can be generated from eco-tourism.
- Limited numbers of people are allowed to visit eco-tourist areas at the one time. For example, daily numbers are limited on the Inca Trail in Peru.

- Tours must be small scale so tour companies like Imaginative Traveller have maximum group sizes ranging from 12 to 24.

- Visitors must follow local customs and respect local cultures, e.g. removing shoes in temples and wearing appropriate clothing.

- Local **guides** educate tourists on the importance of conserving the environment.

Activity 3: Group activity

1. In groups, investigate:
   (a) The cost of an eco-tourist trip to a location of your choice.
   (b) The kit and clothing that you will need to take with you.
   (c) Local projects that you can get involved in.
   (d) How your money is spent to ensure sustainability after you have gone.
2. You must include location maps, graphs and photos to illustrate your findings.
3. You should display the information your group has gathered in a **poster**, a **PowerPoint presentation** or a written **project**.

Summary

In this chapter you have learned:

- Features and causes of mass tourism.

- Features and reasons for eco-tourism.

- Impact of mass tourism and eco-tourism on people and the environment.

- Strategies adopted by governments, conservation agencies and tour companies to manage tourism.

You should have developed your skills and be able to:

- Identify on a map the location of popular tourist and eco-tourist destinations.

- Extract, interpret and present numerical and geographical information which may be statistical, graphical or tabular.

End of chapter questions

National 4 questions

(a) Name **two** features of mass tourism.

(b) Give reasons for the growth in mass tourism.

(c) Name **three** places where mass tourism is popular.

(d) Give **two** positive effects of mass tourism on people.

(e) Give **two** negative effects of mass tourism on the landscape.

(f) How is mass tourism managed by local authorities?

(g) What is eco-tourism?

(h) Give **two** reasons for eco-tourism.

(i) Give **two** positive effects of eco-tourism on the landscape.

(j) Give **two** negative effects of eco-tourism on local people.

National 5 questions

(a) **Describe** the main features of mass tourism.

(b) **Explain** why mass tourism is popular in different places.

(c) **Describe** the effect of mass tourism on local people.

(d) **Discuss** the impact of mass tourism on the natural landscape.

(e) **Explain** ways to reduce the impact of mass tourism in an area you have studied.

(f) **Outline** the key features of eco-tourism.

(g) **Describe** the causes of eco-tourism.

(h) **Discuss** the positive effects that eco-tourism can have on the environment.

(i) **Explain** the negative effects of eco-tourism on people and the natural landscape.

(j) **Describe** the different strategies developed to control the impact of eco-tourism.

National 5 exam-style questions

You can find sample answers to these exam-style questions on the Leckie website:
https://collins.co.uk/pages/scottish-curriculum-free-resources

1   **Describe** the main features of **either** mass tourism **or** eco-tourism.

(4 marks)

2   **Explain** the impact of mass tourism on the environment.
    You should refer to places that you have studied in your answer.

(6 marks)

3

Country	International tourist arrivals (millions)
France	74
USA	55
Spain	52
China	51
Italy	43
UK	28
Turkey	25
Germany	24
Malaysia	23
Mexico	21

Study Diagram Q3 above.

Use the information above to **complete** the bar graph below.

(4 marks)

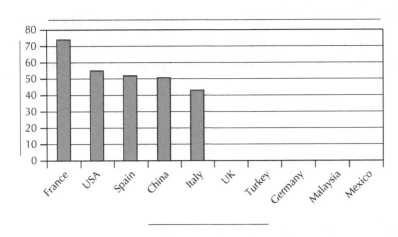

4   **Explain** the importance of eco-tourism for foreign economies.

(6 marks)

LEARNING CHECKLIST

Now that you have finished the **Tourism** chapter, complete a self-evaluation of your knowledge and skills to assess what you have understood. Use traffic lights to help you make up a revision plan to help you improve in the areas you identified as red or amber.

- Locate on a map examples of popular tourist and eco-tourist destinations.

- Outline the features of mass tourism.

- Explain the causes of mass tourism.

- Describe the impact of mass tourism on people.

- Describe the impact of mass tourism on the environment.

- Discuss strategies adopted to manage tourism.

- Outline the features of eco-tourism.

- Explain the reasons for eco-tourism.

- Describe the impact of eco-tourism on people.

- Describe the impact of eco-tourism on the environment.

- Discuss strategies adopted to manage eco-tourism.

Glossary

Blue Flag Scheme: A scheme used to ensure safe, clean beaches and seas for tourists.

Budget airline: A cheap airplane company that offers budget flights.

Conflict: A disagreement.

Conservation: When an area is protected from destruction.

Cultural awareness: When people are aware of the ideas, customs and social behaviour of another society.

Deforestation: The removal of trees from a forest.

Developing country: A country where people have a low standard of living.

Eco-lodge: A house built out of natural resources for tourists to stay in.

Ecosystem: A living community of plants and animals sharing an environment with non-living things such as climate, soil and water.

Eco-tourism: When small groups of people visit remote, untouched environments to support conservation and observe natural vegetation and wildlife.

Eco-tourists: People who travel to remote places to see unspoilt environments.

Environmentally friendly: Activities that are not damaging to the natural environment.

European Union: An organisation consisting of 28 countries joined by an economic and political agreement.

Exploitation: When the natural environment is destroyed for its natural resources.

Future generations: People who will be alive in many years to come.

Global warming: The increase of the Earth's temperature over a period of time.

Globalisation: When views, products and ideas are integrated on a worldwide scale.

Guides: People who know the local area and offer guided tours.

Habitats: The areas where plants and animals live.

Holiday reps: People who are employed in resorts to assist tourists.

Infrastructure: The framework for allowing people to exist in a place, such as roads, houses and schools.

Mass tourism: When many people visit the same place at the same time.

National Park: An area that is protected by law to ensure its conservation.

National Trust: A charity responsible for ensuring the protection and preservation of historic places and spaces for public enjoyment.

Native people: Local people who live in a place.

Nature reserve: A protected area where plants and animals are conserved.

Package deal: A holiday that includes flights, hotel and transfers.

Park ranger: A person who works to protect the natural environment.

Pollution: When the air, water or land are damaged by harmful chemicals.

Preservation: When an area is conserved to ensure it remains unchanged.

Responsible tourism: When visitors are encouraged to cause minimal damage to an area.

Seasonal jobs: Employment which only lasts for part of the year, e.g. during the summer season.

Sewage treatment: An industrial process where contaminants are removed from waste water to ensure safe disposal.

Ski resort: A mountainous area used for skiing that also provides services for tourists.

Social network sites: Internet websites that allow people to communicate, e.g. Facebook.

Soil erosion: When the soil is worn away by people walking over it.

Souvenirs: Objects that tourists buy to remind them of a holiday.

Standard of living: How well off the people of a country are in terms of wealth, health and education.

Stewardship: When people look after the landscape and ensure its protection.

Sustainable: When a resource can be used over and over without it being totally destroyed.

Sustainable tourism: When visitors are encouraged to cause minimal damage to an area to ensure its use for future generations.

Tour operators: A holiday company that combines tours and travel.

Tourism: Travel for recreation or business purposes.

Tourist attractions: Natural environments or man-made structures that attract visitors.

Tourist facilities: Services which are provided for visitors, e.g. restaurants, hotels and souvenir shops.

Tourists: People who travel for recreation or business purposes.

Traditional way of life: The behaviour, habits, ideas and customs that are typical of a particular society.

Traffic congestion: When a large volume of vehicles cause traffic jams.

Travel agent: A company that arranges holidays by booking flights, transfers and hotels for customers.

Vegetation: Trees and plants.

Western cultures: The ideas, customs and social behaviour of the 'western' world, e.g. the USA and Europe.

Wildlife: Non-domesticated/wild plants and animals.